Undergraduate Texts in Mathematics

Editors

S. Axler
F.W. Gehring
P.R. Halmos

Springer
New York
Berlin
Heidelberg
Barcelona
Budapest
Hong Kong
London
Milan
Paris
Santa Clara
Singapore
Tokyo

Undergraduate Texts in Mathematics

(continued after index)

Pierre Brémaud

An Introduction to Probabilistic Modeling

With 90 Illustrations

 Springer

Pierre Brémaud
Laboratoire des Signaux et Systèmes
CNRS
Plateau de Moulon
F91192 Gif-sur-Yvette
France

Editorial Board

S. Axler
Department of Mathematics
Michigan State University
East Lansing, MI 48824
USA

F.W. Gehring
Department of Mathematics
University of Michigan
Ann Arbor, MI 48019
USA

P.R. Halmos
Department of Mathematics
Santa Clara University
Santa Clara, CA 95053
USA

Mathematics Subject Classification (1980): 60-01

Library of Congress Cataloging-in-Publication Data
Brémaud, Pierre.
  An introduction to probabilistic modeling.
  (Undergraduate texts in mathematics)
  Bibliography: p.
  Includes index.
  1. Probabilities. I. Title. II. Series.
QA273.B867   1987   519.2   87-9422

© 1988 by Springer-Verlag New York Inc.
All rights reserved. This work may not be translated or copied in whole or in part without the
written permission of the publisher (Springer-Verlag, 175 Fifth Avenue, New York, New York
10010, USA), except for brief excerpts in connection with reviews or scholarly analysis. Use in
connection with any form of information storage and retrieval, electronic adaptation, computer
software, or by similar or dissimilar methodology now known or hereafter developed is forbidden.
The use of general descriptive names, trade names, trademarks, etc. in this publication, even if
the former are not especially identified, is not to be taken as a sign that such names, as understood
by the Trade Marks and Merchandise Marks Act, may accordingly be used freely by anyone.

Typeset by Asco Trade Typesetting Ltd., Hong Kong.
Printed and bound by R.R. Donnelley and Sons, Harrisonburg, Virginia.
Printed in the United States of America.

9 8 7 6 5 4 3 (Corrected third printing, 1997)

ISBN 0-387-96460-6   Springer-Verlag New York Berlin Heidelberg
ISBN 3-540-96460-6   Springer-Verlag Berlin Heidelberg New York   SPIN 10643494

To my father

Preface

The present textbook provides prerequisite material for courses in Physics, Electrical Engineering, Operations Research, and other fields of applied science where probabilistic models are used intensively. The emphasis has therefore been placed on *modeling* and *computation*.

There are two levels of modeling: abstract and concrete.

The abstract level is relative to the axiomatization of Probability and provides a general framework that features an archetype of all concrete models, where the basic objects (events, probability, random variables), the basic concepts (independence, expectation), and the basic rule (countable additivity of probability) are given in abstract form. This moderately small axiomatic equipment, establishing Probability as a mathematical theory, suffices to produce a theorem called the strong law of large numbers that says in particular that in tossing coins "the average number of heads tends to $\frac{1}{2}$ as the number of independent tosses tends to infinity, if the coin is fair." This result shows that the axioms of probability are consistent with empirical evidence. (From a mathematical point of view, this a posteriori check of the relevance of the axioms is not necessary, whereas from the point of view of the modeler, it is of course of paramount importance.)

In the present book, the abstract framework is immediately introduced and a number of examples showing how this framework relates to the daily concerns of physicists and engineers is provided. The strong law of large numbers where the abstract framework culminates is proved in the last chapter.

The other level of modeling consists of fitting a given situation into the conceptual framework of the axiomatic theory when it is believed that random phenomena occur. This is a difficult exercise at the beginning, and the art of modeling can be acquired only through examples. Supplementary readings—

entitled Illustrations— provide examples in which probabilistic models have
been successfully developed.

They include, in particular, topics in *stochastic processes* and *statistics* as
shown in the following list:

1. A Simple Model in Genetics: Mendel's Law and Hardy–Weinberg's
   Theorem
2. The Art of Counting: The Ballot Problem and the Reflection Principle
3. Bertrand's Paradox
4. An Introduction to Population Theory: Galton–Watson's Branching
   Process
5. Shannon's Source Coding Theorem: An Introduction to Information
   Theory
6. Buffon's Needle: A Problem in Random Geometry
7. An Introduction to Bayesian Decision Theory: Tests of Gaussian Hypotheses
8. A Statistical Procedure: The Chi-Square Test
9. Introduction to Signal Theory: Filtering.

The first chapter introduces the basic definitions and concepts of probability,
independence, and cumulative distribution functions. It gives the elementary
theory of conditioning (Bayes' formulas), and presents finite models, where
computation of probability amounts to counting the elements of a given
set. The second chapter is devoted to *discrete random variables* and to the
generating functions of integer-valued random variables, whereas the third
chapter treats the case of *random vectors admitting a probability density*. The
last paragraph of the third chapter shows how Measure and Integration
Theory can be useful to Probability Theory. It is of course just a brief summary
of material far beyond the scope of an introduction to probability, emphasiz-
ing a useful technical tool: the Lebesgue convergence theorems. The fourth
chapter treats two topics of special interest to engineers, operations researchers,
and physicists: the *Gaussian vectors* and the *Poisson process*, which are the
building blocks of a large number of probabilistic models. The treatment
of Gaussian vectors is elementary but nevertheless contains the proof of
the stability of the Gaussian character by extended linear transformations
(linear transformations followed by passage to the limit in the quadratic
mean). The Gaussian vectors and the Poisson process also constitute a source
of examples of application of the formula of transformation of probability
densities by smooth transformations of random vectors, which is given in
the first paragraph and provides unity for this chapter. The last chapter
treats the various concepts of *convergence*: in probability, almost sure, in
distribution, and in the quadratic mean.

About 120 exercises with detailed solutions are presented in the main text
to help the reader acquire computational skills and 28 additional exercises
with outlines of solutions are given at the end of the book.

The material of the present textbook can be covered in a one-semester undergraduate course and the level can be adjusted simply by including or discarding portions of the last chapter, more technical, on convergences. The mathematical background consists of elementary calculus (series, Riemann integrals) and elementary linear algebra (matrices) as required of students in Physics and Engineering departments.

Gif-sur-Yvette, France   PIERRE BRÉMAUD

Contents

CHAPTER 3
Probability Densities 85

CHAPTER 4
Gauss and Poisson 128

Abbreviations and Notations

Abbreviations

a.s.	almost surely
c.d.f.	cumulative distribution function
c.f.	characteristic function
i.i.d.	independent and identically distributed
p.d.	probability density
q.m.	quadratic mean
r.v.	random variable

Notations

$\mathscr{B}(n, p)$	the binomial law of size n and parameter p (p. 47)
$\mathscr{G}(p)$	the geometric law of parameter p (p. 48)
$\mathscr{P}(\lambda)$	the Poisson law of mean λ (p. 49)
$\mathscr{M}(n, k, p_i)$	the multinomial law of size (n, k) and parameter $(p_1, \ldots, p_k)$ (p. 49)
$\mathscr{U}([a, b])$	the uniform law over $[a, b]$ (p. 86)
$\mathscr{E}(\lambda)$	the exponential law of parameter λ (p. 86)
$\mathscr{N}(m, \sigma^2)$	the Gaussian law of mean m and variance σ^2 (p. 87)
$\gamma(\alpha, \beta)$	the gamma law of parameters α and β (p. 88)
χ_n^2	the chi-square law with n degrees of freedom (p. 88)
$X \sim \cdots$	the random variable X is distributed according to ... (Example: "$X \sim \mathscr{E}(\lambda)$" means "$X$ is distributed according to the exponential law of parameter λ)
$\mathbb{R}$	the set of real numbers
$\mathbb{R}^n$	the set of n-dimensional real vectors

$\mathscr{B}^n$	the Borelian sets of $\mathbb{R}^n$, that is: the smallest σ-field on $\mathbb{R}^n$ containing all the n-dimensional "rectangles"
$\mathbb{N}$	the set of non-negative integers
A'	transpose of the matrix A
u'	line vector, transpose of the column vector u

CHAPTER 1

Basic Concepts and Elementary Models

1. The Vocabulary of Probability Theory

In Probability Theory the basic object is a probability space $(\Omega, \mathscr{F}, P)$ where

Ω is the collection of all possible *outcomes* of a given experiment;

$\mathscr{F}$ is a family of subsets of Ω, called the family of *events*; and

P is a function from $\mathscr{F}$ into $[0, 1]$ assigning to each event $A \in \mathscr{F}$ its *probability* $P(A)$.

The mathematical objects $\mathscr{F}$ and P must satisfy a few requirements, called the *axioms of Probability Theory*, which will be presented in Section 2.1.

Although it is quite acceptable from a mathematical point of view to present the axioms of Probability without attempting to interpret them in terms of daily life, it is preferable to start by showing what reality they are supposed to symbolize and to give the formal definitions later. In this respect, a small lexicon of the terms used by probabilists will be useful.

Trial. The confusion between an experimental setting and actual trials is often made. The term *experimental setting* refers to the general conditions under which various trials are performed. For instance, the Michelson–Morley experimental setting consists of a method (interferometry) and of an apparatus for measuring very small relative variations in the velocity of light. Michelson and Morley performed several trials in this experimental setting. In Statistics, the experimental setting consists of the conditions under which data are collected, and a trial might consist, for instance, of the actual conduct of a poll.

Outcome, Sample. In Probability, one considers trials performed in a given experimental setting. Any experiment has an outcome, or result. It has been a great conceptual advance to abandon the idea of capturing the notion of experiment in mathematical terms; experiment, like experience, is too deeply rooted in the physical and psychological world. Instead the mathematicians have chosen to consider the *outcome* because it is more amenable to mathematics.

Indeed, only a writer can convincingly describe the wicked "coup de main" of the croupier, the whimsical motion of the ball on the wheel, and the ensuing torments in the player's soul. For conciseness, a mathematician prefers to deal with the possible outcomes: 0, 1, ..., 37. Another term for outcome is *sample*.

> The *sample space* Ω is the collection of all possible outcomes ω.

An Event. For an experiment being performed and its outcome ω being observed, one can tell whether such and such an event has occurred. An event is best described by a subset of the sample space. For instance, in the game of roulette the set of outcomes $\{0, 2, 4, \ldots, 36\}$ is an event. One can use a picturesque name for this event, such as "even," or any other name, depending on one's imagination. But an event is nevertheless just a collection of outcomes, i.e., a subset $A \subset \Omega$.

> An *event* is a collection of outcomes, i.e., a subset A of the sample space Ω.
> If $\omega \in A$, one says that *outcome ω realizes event A.*

This is a temporary definition; the complete definition will be given in Section 2.1.

The Logics of Events. If subset A is contained in subset $B (A \subset B)$, this is expressed by *event A implies B* (Fig. 1). Two events A and B are *incompatible* when there exists no outcome ω that realizes both A and B, i.e., $A \cap B = \emptyset$, where $\emptyset$ is the empty set. Consider now a family of events $A_1, \ldots, A_k$. The set equality $\Omega = \bigcup_{i=1}^{k} A_i$ means that at least one of the events $A_1, \ldots, A_k$ is realized. Indeed either $\omega \in A_1$, or $\omega \in A_2, \ldots$, or $\omega \in A_k$. It is clear now that any relation or equality between subsets of Ω is the formal transcription of a logical relation between events. Another example is the set equality $\Omega = \sum_{i=1}^{k} A_i$ (i.e., $\Omega = \bigcup_{i=1}^{k} A_i$ and $A_i \cap A_j = \emptyset$ when $i \neq j$), which tells that the events $A_1, \ldots, A_k$ are *exhaustive and mutually incompatible*. In other words, one and only one among the events $A_1, \ldots, A_k$ will happen. When event A is not realized, event $\bar{A}$ is realized where $\bar{A}$ is the complement of A. For obvious reasons, Ω is called the *certain* event and $\emptyset$, the *impossible* event.

It is now time for the formal definition of a probability space, which will be given in Section 2.1, followed by illustrative examples in Section 2.2.

Picture language	Set theoretical notation	Logical meaning in terms of events
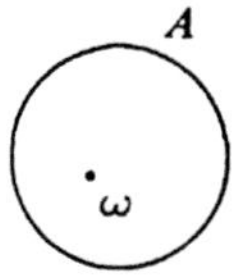	$\omega \in A$	ω realizes A
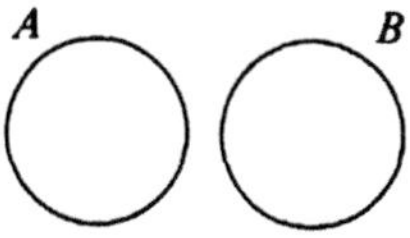	$A \cap B = \phi$	A and B are incompatible
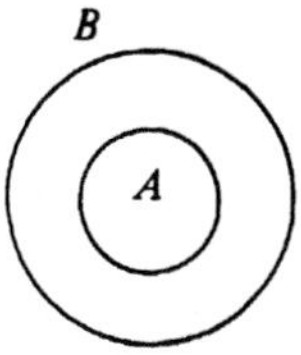	$A \subset B$	A implies B
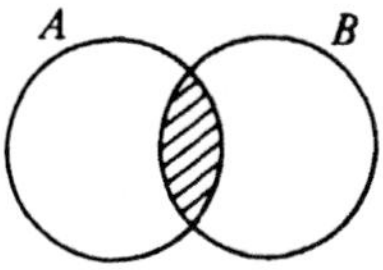	$A \cap B$	A and B are both realized
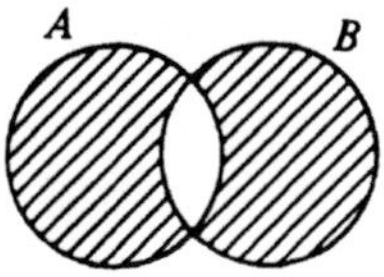	$A \triangle B$	One and only one of the events A and B is realized
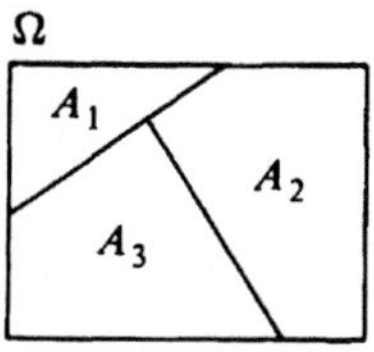	$\Omega = A_1 + A_2 + A_3$	One and only one of the events A_1, A_2, A_3 is realized by any sample ω.

Figure 1. A probabilist's view of sets. Note that the symbol $\sum$ can be used in place of $\bigcup$ only if the sets in the union are disjoint.

2. Events and Probability

2.1. Probability Space

A probabilistic model (or *probability space*) consists of a triple $(\Omega, \mathcal{F}, P)$ where

Ω is the *sample space*, a collection of outcomes ω.

$\mathcal{F}$ is a collection of subsets of Ω; a subset $A \in \mathcal{F}$ is called an *event*.

P is a set function mapping $\mathcal{F}$ into the interval $[0, 1]$: with each event $A \in \mathcal{F}$, it associates the *probability* $P(A)$ of this event.

The collection $\mathcal{F}$ and the mapping P are required to satisfy the following axioms.

Axioms Relative to the Events. The family $\mathcal{F}$ of events must be a *σ-field* on Ω, that is,

(i) $\Omega \in \mathcal{F}$

(ii) if $A \in \mathcal{F}$, then $\bar{A} \in \mathcal{F}$ (where $\bar{A}$ is the complement of A)

(iii) if the sequence $(A_n, n \geqslant 1)$ has all its members in $\mathcal{F}$, then the union $\bigcup_{n=1}^{\infty} A_n \in \mathcal{F}$.

Axioms Relative to the Probability. The probability P is a mapping from $\mathcal{F}$ into $[0, 1]$ such that:

(α) $P(\Omega) = 1$

(β) for any sequence $(A_n, n \geqslant 1)$ of *disjoint* events of $\mathcal{F}$, the following property, called *σ-additivity*, holds:

$$\boxed{P\left(\sum_{n=1}^{\infty} A_n \right) = \sum_{n=1}^{\infty} P(A_n)}. \tag{1}$$

An event A such that $P(A) = 1$ is called an *almost certain* event. Similarly, if $P(A) = 0$, A is called an *almost impossible* event. Two events A and B such that $P(A \cap B) = 0$ are said to be *probabilistically incompatible*.

Note that $A = \varnothing$ implies from the axioms [see Eq. (3)] that $P(A) = 0$, but the converse is not true: an event can be logically possible and, nevertheless, have no chance of happening.

Immediate Properties of Probability. A few properties follow directly from the axioms. First, since $\Omega \in \mathcal{F}$, its complement $\bar{\Omega} = \varnothing$ also lies in $\mathcal{F}$, by (ii). Second, if $(A_n, n \geqslant 1)$ is a sequence of events, then the intersection $\bigcap_{n=1}^{\infty} A_n$ is also an event. The latter assertion is proven by applying de Morgan's formula

$$\bigcap_{n=1}^{\infty} A_n = \left(\overline{\bigcup_{n=1}^{\infty} \bar{A}_n} \right),$$

and using successively axioms (ii), (iii), and (ii).

From axioms (α) and (β) we obtain

$$\boxed{P(\bar{A}) = 1 - P(A)}. \tag{2}$$

Indeed, $1 = P(\Omega) = P(A + \bar{A}) = P(A) + P(\bar{A})$. By specializing this equality to $A = \Omega$, we have $P(\emptyset) = 0$. Summarizing this result and previous relations from the set of axioms, we have for any event A

$$\boxed{P(\emptyset) = 0 \leqslant P(A) \leqslant 1 = P(\Omega)}. \tag{3}$$

If A logically implies B, i.e., $A \subset B$, the set $B - A$ is well defined and $A + (B - A) = B$. Therefore, by the σ-additivity axiom, $P(A) + P(B - A) = P(B)$, or

$$\boxed{A \subset B \Rightarrow P(B - A) = P(B) - P(A)}. \tag{4}$$

In particular, since $P(B - A) \geqslant 0$, the mapping P is monotone increasing:

$$\boxed{A \subset B \Rightarrow P(A) \leqslant P(B)}. \tag{5}$$

E1 Exercise. Let $(A_n, n \geqslant 1)$ be an arbitrary sequence of events. Prove the following property, called *sub-σ-additivity*:

$$\boxed{P\left(\bigcup_{n=1}^{\infty} A_n\right) \leqslant \sum_{n=1}^{\infty} P(A_n)}. \tag{6}$$

Hint: Use the set identity $\bigcup_{n=1}^{\infty} A_n = \sum_{n=1}^{\infty} A'_n$, where $A'_1 = A_1$, $A'_n = A_n - A_n \cap (\bigcup_{j=1}^{n-1} A_j)$ for $n \geqslant 2$ (Fig. 2).

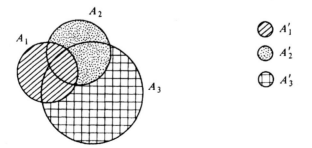

Figure 2.

E2 Exercise. Let A, B, C be three arbitrary events. Show that

$$P(A \cup B) = P(A) + P(B) - P(A \cap B),$$

and

$$P(A \cup B \cup C) = P(A) + P(B) + P(C) - P(A \cap B) - P(A \cap C)$$
$$- P(B \cap C) + P(A \cap B \cap C).$$

The above formulas are particular instances of a general formula, called the *inclusion–exclusion formula*, which will be given in Eq. (29) of Chapter 2.

We shall now give a very useful consequence of the σ-additivity axiom.

Sequential Continuity of Probability. Consider a sequence of events $(B_n, n \geqslant 1)$ such that

$$\boxed{B_n \subset B_{n+1} \qquad (n \geqslant 1)} \tag{7}$$

Then

$$\boxed{P\left(\bigcup_{n=1}^{\infty} B_n\right) = \lim_{n \uparrow \infty} \uparrow P(B_n)} \tag{8}$$

This property is the *sequential continuity of probability* because $\bigcup_{n=1}^{\infty} B_n$ is called the (increasing) limit of the (increasing) sequence $(B_n, n \geqslant 1)$ and is denoted $\lim_{n \uparrow \infty} \uparrow B_n$, so that Eq. (8) reads $P(\lim_{n \uparrow \infty} \uparrow B_n) = \lim_{n \uparrow \infty} \uparrow P(B_n)$.

PROOF OF EQ. (8). Observe that

$$B_p = \bigcup_{n=1}^{p} B_n = B_1 + \sum_{n=1}^{p-1} (B_{n+1} - B_n)$$

and

$$\bigcup_{n=1}^{\infty} B_n = B_1 + \sum_{n=1}^{\infty} (B_{n+1} - B_n).$$

Therefore, (σ-additivity)

$$P\left(\bigcup_{n=1}^{\infty} B_n\right) = P(B_1) + \sum_{n=1}^{\infty} P(B_{n+1} - B_n)$$

$$= \lim_{p \uparrow \infty} \left(P(B_1) + \sum_{n=1}^{p-1} P(B_{n+1} - B_n)\right)$$

$$= \lim_{p \uparrow \infty} P(B_p).$$

Similarly, if $(C_n, n \geq 1)$ is a decreasing sequence of events, i.e.,

$$\boxed{C_{n+1} \subset C_n \qquad (n \geq 1)}, \tag{9}$$

then

$$\boxed{P\left(\bigcap_{n=1}^{\infty} C_n\right) = \lim_{n \uparrow \infty} \downarrow P(C_n)}. \tag{10}$$

$\square$

PROOF OF EQ. (10). Apply the previous result to $B_n = \bar{C}_n$. By de Morgan's rule and property (2), $P(\bigcap_{n=1}^{\infty} C_n) = P(\overline{\bigcup_{n=1}^{\infty} \bar{C}_n}) = 1 - P(\bigcup_{n=1}^{\infty} \bar{C}_n) = 1 - \lim_{n \uparrow \infty} P(\bar{C}_n) = 1 - \lim_{n \uparrow \infty} (1 - P(C_n)) = \lim_{n \uparrow \infty} P(C_n)$. $\square$

2.2. Two Elementary Probabilistic Models

Before proceeding further into the examination of the consequences of the probability axioms, we will give two examples of probability models.

EXAMPLE 1 (Choosing a Point at Random in the Unit Square). Here the point will be supposed to be "completely randomly" chosen in $[0, 1] \times [0, 1]$. The following model is proposed.

First, $\Omega = [0, 1] \times [0, 1]$, that is to say: any *outcome* ω has the form $\omega = (x, y)$, where $0 \leq x \leq 1$ and $0 \leq y \leq 1$ (Fig. 3).

We will be rather vague in the description of the collection $\mathscr{F}$ of events, calling an event any subset A of $\Omega = [0, 1] \times [0, 1]$ for which one can define the area $S(A)$. For instance, any set A of the form $A = [x_1, x_2] \times [y_1, y_2]$, in which case $S(A) = (x_2 - x_1)(y_2 - y_1)$. There are many more sets in $\mathscr{F}$. However, the description of some of them is not easy. This matter will be discussed later since it is of minor interest at this stage.

The probability $P(A)$ is just the area of A, $S(A)$. The mapping $A \to S(A)$ is indeed a mapping from $\mathscr{F}$ into $[0, 1]$, and the first axiom (α) is satisfied since

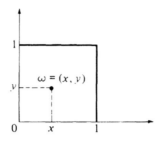

Figure 3. The sample space Ω relative to the experiment of choosing at random a point in the unit square.

$P(\Omega) = S([0,1]^2) = 1$. The axiom of σ-additivity is also satisfied, in accordance with the intuitive notion of area. But we are not able to prove this formally here, having not yet properly defined the σ-field of events $\mathscr{F}$.

EXAMPLE 2 (Three-Dice Game). The probabilistic description of a roll of three dice can be made as follows. The sample space Ω consists of the collection of all triples $\omega = (i, j, k)$ where i, j, k are integers from 1 to 6. In abbreviated notations, $\Omega = \{1, 2, 3, 4, 5, 6\}^3$. The dice are supposed to be distinguishable, and i is the outcome of the first die and j and k are the outcomes of the second and third dice, respectively (Fig. 4).

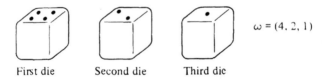

First die   Second die   Third die

Figure 4. The event "odd sum" is realized by this particular outcome.

The choice of $\mathscr{F}$ is the trivial one: $\mathscr{F}$ is the collection of *all* subsets of Ω (it is obviously a σ-field). As for P, it is defined by $P(\{\omega\}) = (\frac{1}{6})^3$ for all ω. Here $\{\omega\}$ is the set consisting of the unique element ω. This suffices to define P for all subsets A of Ω since by the σ-additivity axiom we must have $P(A) = \sum_{\omega \in A} P(\{\omega\})$, this quantity being taken by convention to be 0 if A is empty. In summary,

$$P(A) = (\tfrac{1}{6})^3 \cdot |A|,$$

where $|A|$ is the cardinality of A, that is, the number of elements in A. From this, σ-additivity is obvious and $P(\Omega) = 1$ since Ω contains 6^3 elements.

This type of model where the probability of an event is proportional to its cardinality will be considered in more detail in Section 6 of the present chapter.

3. Random Variables and Their Distributions

3.1. Random Variables (r.v.)

The single most important notion of Probability Theory is that of random variable. A random variable on $(\Omega, \mathcal{F})$ is any mapping $X: \Omega \to \bar{\mathbb{R}}$ such that for all $a \in \mathbb{R}$,

$$\boxed{\{\omega | X(\omega) \leqslant a\} \in \mathcal{F}} \ . \tag{11}$$

Here $\mathbb{R}$ is the set of real numbers, and $\bar{\mathbb{R}}$ is the extension of $\mathbb{R}$ consisting of $\mathbb{R}$ plus the two infinite numbers $+\infty$ and $-\infty$. The notation $\{\omega | X(\omega) \leqslant a\}$ represents the collection of ω's such that $X(\omega) \leqslant a$. It will usually be abbreviated as $\{X \leqslant a\}$.

Requirement (11) ensures that one can compute the probability of events $\{X \leqslant a\}$, since the probability P is defined only for subsets of Ω belonging to $\mathcal{F}$.

EXAMPLE 3. The setting is as in Example 1, where all outcomes have the form $\omega = (x, y)$. One can define two mappings X and Y from Ω into $[0, 1]$ by

$$X(\omega) = x, \qquad Y(\omega) = y.$$

Both X and Y are random variables since in the case of X, for instance, and when $a \in [0, 1]$, the set $\{X \leqslant a\}$ is the rectangle $[0, a] \times [0, 1]$, a member of $\mathcal{F}$. For $a \geqslant 1$, $\{X \leqslant a\} = \Omega$, and for $a < 0$, $\{X \leqslant a\} = \varnothing$, so that Eq. (11) is also verified.

E3 Exercise. In the setting of Example 3, show that the mapping $Z = X + Y$ (i.e., $Z(\omega) = X(\omega) + Y(\omega)$) is a random variable. Compute for all $a \in \mathbb{R}$ the probability of event $\{Z \leqslant a\}$.

EXAMPLE 4. Consider the three-dice game of Example 2. Here an outcome ω has the form $\omega = (i, j, k)$. We can define three mappings X_1, X_2, X_3 from Ω into $\{1, 2, 3, 4, 5, 6\}$ by

$$X_1(\omega) = i, \qquad X_2(\omega) = j, \qquad X_3(\omega) = k.$$

The verification that X_1, X_2, X_3 are random variables is immediate since, in this particular probabilistic model, $\mathcal{F}$ contains all subsets of Ω.

3.2. Cumulative Distribution Function (c.d.f.)

The cumulative distribution function of the random variable X is the function F mapping $\mathbb{R}$ into $[0, 1]$ defined by

$$F(x) = P(X \leqslant x) \quad . \tag{12}$$

The notation $P(X \leqslant x)$ is an abbreviation of $P(\{X \leqslant x\})$.

A cumulative distribution function F is *monotone increasing* since whenever $x_1 \leqslant x_2$, $\{X \leqslant x_1\} \subset \{X \leqslant x_2\}$ and therefore, by the monotonicity property (5), $P(X \leqslant x_1) \leqslant P(X \leqslant x_2)$.

We will see below that F is a *right-continuous* function. This property depends very much on the "less than or equal to" sign in Eq. (12). Had we chosen to define $F(x)$ as $P(X < x)$ with a "less than" sign, F would have been left-continuous. (This convention is seldom adopted.)

Any right-continuous increasing function F admits a left-hand limit at all points $x \in \mathbb{R}$, denoted by $F(x-)$. We will see in a few lines that if we define

$$F(+\infty) = \lim_{x \uparrow +\infty} F(x), \qquad F(-\infty) = \lim_{x \downarrow -\infty} F(x), \tag{13}$$

then

$$1 - F(+\infty) = P(X = +\infty), \qquad F(-\infty) = P(X = -\infty) \quad . \tag{14}$$

PROOF OF THE RIGHT CONTINUITY OF F. Let $x \in \mathbb{R}$ and let $(\varepsilon_n, n \geqslant 1)$ be a sequence of strictly positive real numbers decreasing to 0. Define for each $n \geqslant 1$, $C_n = \{X \leqslant x + \varepsilon_n\}$. Then $(C_n, n \geqslant 1)$ is a decreasing sequence of events and $\bigcap_{n=1}^{\infty} C_n = \{X \leqslant x\}$. By Eq. (10), $P(X \leqslant x) = \lim_{n \uparrow \infty} \downarrow P(X \leqslant x + \varepsilon_n)$, i.e., $F(x) = \lim_{n \uparrow \infty} \downarrow F(x + \varepsilon_n)$, qed. $\square$

PROOF OF EQ. (14). Define $B_n = \{X \leqslant n\}$ and $C_n = \{X \leqslant -n\}$. Then $\bigcup_{n=1}^{\infty} B_n = \{X < \infty\}$ and $\bigcap_{n=1}^{\infty} C_n = \{X = -\infty\}$, and Eq. (14) follows from Eqs. (8) and (10). $\square$

EXAMPLE 5. We consider the probabilistic model of Examples 1 and 3 (choosing a point at random in the unit square). The cumulative distribution function of X has the graph shown in Fig. 5. Indeed, if $x \geqslant 1$, then $\{\omega | X(\omega) \leqslant x\} = \Omega$ and therefore $P(X \leqslant x) = P(\Omega) = 1$, and if $x < 0$, then $\{\omega | X(\omega) \leqslant x\} = \varnothing$ and therefore $P(X \leqslant x) = P(\varnothing) = 0$. Also, when $x \in [0, 1]$, the set $\{X \leqslant x\}$ is the rectangle $[0, x] \times [0, 1]$ of area x.

Random Variables with a Probability Density (p.d.). In the general case, a random variable X may take the values $+\infty$ and/or $-\infty$. If it takes only finite values, X is called a *real random variable*.

If a *real* random variable X admits a cumulative distribution function F such that

$$F(x) = \int_{-\infty}^{x} f(y) \, dy \tag{15}$$

for some nonnegative function f, then X is said to admit the *probability density* f. It must be pointed out that f then satisfies

$$\int_{-\infty}^{+\infty} f(y)\,dy = 1 \;. \tag{16}$$

The above equality follows from Eqs. (15) and (14).

EXAMPLE 6 (Continuation of Example 5). The random variable X with the cumulative distribution function F of Fig. 5 admits a probability density f (Fig. 6) where

$$f(x) = \begin{cases} 0 & \text{if } x < 0 \\ 1 & \text{if } x \in [0, 1] \\ 0 & \text{if } x > 1. \end{cases}$$

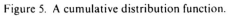

Figure 5. A cumulative distribution function.

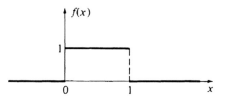

Figure 6. The uniform density over $[0, 1]$.

The distribution pictured in Fig. 5 or 6 is the *uniform distribution over* $[0, 1]$. The corresponding random variable is said to be *uniformly distributed over* $[0, 1]$.

E4 Exercise. Find the c.d.f. and the p.d. of the random variable defined in Exercise E3.

Discrete Random Variables. Let X be a random variable taking only integer values, i.e., values in the set $\mathbb{N} = \{0, 1, 2, \ldots\}$, and denote $p_n = P(X = n)$. In this special case, since $\{X \leqslant x\} = \sum_{n, n \leqslant x} \{X = n\}$, we have by the σ-additivity axiom the following expression for the c.d.f.:

$$F(x) = \sum_{n, n \leqslant x} p_n.$$

The function F is then purely discontinuous, as shown in Fig. 7.

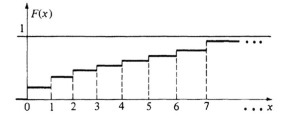

Figure 7. The c.d.f of an integer valued random variable. The jump of F at $x = n$ is of magnitude p_n.

A random variable X taking only a denumerable set of values $(a_n, n \geqslant 0)$ is called a *discrete random variable*. Its distribution is the sequence $(p_n, n \geqslant 0)$ where

$$\boxed{p_n = P(X = a_n)}. \tag{17}$$

The numbers p_n are nonnegative, and they verify

$$\boxed{\sum_{n=0}^{\infty} p_n = 1}. \tag{18}$$

This relation is obtained from $\Omega = \sum_{n=0}^{\infty} \{X = a_n\}$ and the σ-additivity axiom.

E5 Exercise. Consider the three-dice game of Examples 2 and 4 and the discrete random variable $X = X_1 + X_2 + X_3$. Compute $p_{18} = P(X = 18)$ and $p_6 = P(X = 6)$.

4. Conditional Probability and Independence

4.1. Independence of Events

Conditional Probability. Let B be an event of *strictly positive* probability. For any event A, one defines the symbol $P(A|B)$ by

$$P(A|B) = \frac{P(A \cap B)}{P(B)}. \tag{19}$$

The quantity $P(A|B)$ is called the *probability of A given B*. It admits an interesting interpretation which makes it one of the fundamental concepts of Probability Theory. Suppose that n experiments have been performed and that the occurrences of events A and B have been recorded as follows: n_A among the n experiments have resulted in the realization of event A, n_B in the realization of B, and $n_{A \cap B}$ in the joint realization of A and B. The *frequency* interpretation of probability, which will be given a firm mathematical basis with the *law of large numbers*, suggests that if all experiments have been performed "independently," the frequency n_A/n is close to the probability $P(A)$ when n is large. Similar statements hold for $P(B)$ and $P(A \cap B)$, so that the quantity $P(A|B)$ is by Eq. (19) close to $n_{A \cap B}/n$: $n_B/n = n_{A \cap B}/n_B$. Now $n_{A \cap B}/n_B$ is the relative frequency of A among the realizations of B. Just as n_A/n measured our expectation of observing event A, $n_{A \cap B}/n_B$ measures our expectation of seeing A realized knowing that B is realized.

As an illustration, imagine that a sample of $n = 10,000$ individuals have been selected at random in a given Irish town. It is observed that among them, $n_A = 5,000$ have blue eyes, $n_B = 5,000$ have black hair, and $n_{A \cap B} = 500$ have blue eyes and black hair. From these data, we must expect that, with a probability approximately equal to $n_A/n = \frac{1}{2}$, the first person to be met in the street has blue eyes. But what if this citizen wears dark glasses hiding his eyes and has black hair? We then expect that his eyes are blue with a probability approximately equal to $n_{A \cap B}/n_B = 1/10$. What we have done here is to replace the "a priori" probability $P(A)$, by the "conditional" probability of A given B, $P(A|B)$.

Independence. Let us continue with the above population sample and suppose that $n_C = 1,000$ citizens among the $n = 10,000$ have a name starting with one of the first seven letters of the alphabet. We have the feeling that eye color and initials are "independent." For this reason, we believe that the proportion of blue eyes among the citizens with a name beginning with one of the first seven letters of the alphabet is approximately the same as the proportion of blue eyes in the whole population, i.e., $n_A/n \simeq n_{A \cap C}/n_C$ or, in the probabilistic idealization, $P(A) = P(A|C)$. Now, from the definition of $P(A|C)$, the latter relation is simply $P(A \cap C) = P(A)P(C)$. One is therefore led to adopt the following definition: two events A and B are said to be *independent* iff

$$P(A \cap B) = P(A)P(B). \tag{20}$$

E6 Exercise. Consider the following table:

	$P(A)$	$P(B)$	$P(A \cup B)$
Case 1	0.1	0.9	0.91
Case 2	0.4	0.6	0.76
Case 3	0.5	0.3	0.73

For which cases are events A and B independent?

The definition of independence extends to the case of an arbitrary family of events as follows. Let $\mathscr{C}$ be an arbitrary family of events (finite, countable or uncountable). This family $\mathscr{C}$ is said to be a family of independent events iff, for *all finite* subfamilies $\{A_1, \ldots, A_n\}$ of $\mathscr{C}$,

$$P\left(\bigcap_{j=1}^{n} A_j \right) = \prod_{j=1}^{n} P(A_j). \tag{21}$$

E7 Exercise. Let $\{A_1, \ldots, A_k\}$ be a family of independent events. Show that $\{\bar{A}_1, A_2, \ldots, A_k\}$ is also a family of independent events.

The above exercise shows that in an arbitrary family of independent events, one can replace an arbitrary number (finite, countable, or uncountable) of events by their complement and still retain the independence property for the resulting family.

The next exercise points out a beginner's mistake.

E8 Exercise. Let $\Omega = \{\omega_1, \omega_2, \omega_3, \omega_4\}$ be a sample space with just four points, and $\mathscr{F}$ be the family of all subsets of Ω. A probability P is defined on $\mathscr{F}$ by $P(\{\omega_i\}) = \frac{1}{4}(1 \leqslant i \leqslant 4)$. Let A, B, C be the following events:

$$A = \{\omega_1, \omega_2\}, \qquad B = \{\omega_2, \omega_3\}, \qquad C = \{\omega_3, \omega_1\}.$$

Show that $\mathscr{C} = \{A, B, C\}$ is not a family of independent events, although A is independent of B, A is independent of C, and B is independent of C.

Remark. Another beginner's mistake: disjoint events are independent. This is *wrong*. Indeed, if it were true, then for every pair of disjoint events A and B, at least one of them would be of probability zero, in view of $0 = P(\varnothing) = P(A \cap B) = P(A)P(B)$. As a matter of fact, two disjoint events are strongly dependent since "disjoint" means "incompatible": if one of them is realized then you know that the other is not.

Conditional Independence. Let C be an event of strictly positive probability. Define P_C, a mapping from the family of events $\mathscr{F}$ into $\mathbb{R}$, by

$$P_C(A) = P(A|C)$$
(22)

E9 Exercise. Show that P_C is a probability on $(\Omega, \mathscr{F})$.

Two events A and B that are independent relatively to probability P_C are said to be *conditionnally independent given* C. The defining formula is

$$P(A \cap B|C) = P(A|C)P(B|C)$$
(23)

This fundamental concept of Probability Theory will be illustrated by Exercise E14 of Section 5.

E10 Exercise. Let C and D be two events such that $P(C \cap D) > 0$. Verify that for any event A

$$P_C(A|D) = P(A|C, D)$$

where $P(A|C, D) = P(A|C \cap D)$.

4.2. Independence of Random Variables

The concept of independence of events extends to random variables in a natural way. Two random variables X and Y defined on $(\Omega, \mathscr{F}, P)$ are said to be independent iff for all $a, b \in \mathbb{R}$,

$$P(X \le a, Y \le b) = P(X \le a)P(Y \le b)$$
(24)

Here the notation $P(X \le a, Y \le b)$ is an abbreviation of $P(\{X \le a\} \cap \{Y \le b\})$.

EXAMPLE 7. The two random variables X and Y defined in Example 3 are independent. Indeed, for a and b in $[0, 1]$, for instance, $P(X \le a, Y \le b)$ is the area of $[0, a] \times [0, b]$, i.e., ab, and $P(X \le a) = a$, $P(Y \le b) = b$ (see Fig. 8).

In the case where X and Y are discrete random variables taking the values $(a_n, n \ge 0)$ and $(b_m, m \ge 0)$, respectively, requirement (24) for all a and b is equivalent to the requirement that for all $n \ge 0, m \ge 0$,

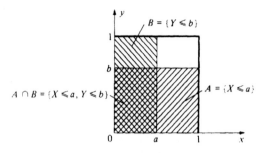

Figure 8. The unit square and its coordinate random variables.

$$P(X = a_n, Y = b_m) = P(X = a_n)P(Y = b_m). \tag{25}$$

The verification of this assertion is left to the reader.

The extension to several random variables of the notion of independence is straightforward.

A family $\mathscr{H}$ of random variables is said to be independent if for any finite subfamily $\{Y_1, \ldots, Y_n\} \subset \mathscr{H}$ the relation

$$P(Y_1 \leq a_1, \ldots, Y_n \leq a_n) = \prod_{j=1}^{n} P(Y_j \leq a_j), \tag{26}$$

holds for all $a_j \in \mathbb{R}(1 \leq j \leq n)$.

In the case of discrete random variables, a simpler definition is available: just replace Eq. (26) with

$$P(Y_1 = a_1, \ldots, Y_n = a_n) = \prod_{j=1}^{n} P(Y_j = a_j), \tag{27}$$

where a_j ranges over the set of values of Y_j.

E11 Exercise (The Binomial Distribution). Let $X_1, \ldots, X_n$ be n discrete random variables taking their values in $\{0, 1\}$ and with the same distribution

$$P(X_j = 1) = p, P(X_j = 0) = 1 - p. \tag{28}$$

Suppose, moreover, that they are *independent*. Defining

$$S_n = X_1 + \cdots + X_n \tag{29}$$

(a random variable taking integer values from 0 to n), show that

$$P(S_n = k) = \frac{n!}{k!(n-k)!} p^k (1-p)^{n-k}. \quad (0 \leqslant k \leqslant n)$$ (30)

The probability distribution $(p_k, 0 \leqslant k \leqslant n)$ given by

$$p_k = \frac{n!}{k!(n-k)!} p^k (1-p)^{n-k}$$ (31)

is called the *binomial distribution* of size n and parameter p (see Fig. 9). Any discrete random variable X admitting this distribution is called a *binomial random variable* (of size n and parameter p). This is denoted by $X \sim \mathscr{B}(n, p)$.

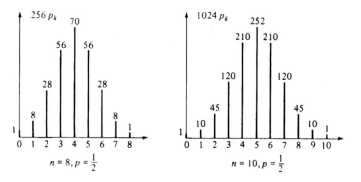

Figure 9. The binomial law.

5. Solving Elementary Problems

5.1. More Formulas

This section is devoted to the illustration of the notions and concepts previously introduced. However, before proceeding to the statement and solution of a few classic exercises, we will start a collection of simple formulas of constant use in this type of exercise (and in many other situations).

Bayes' Retrodiction Formula. The data of a given statistical problem sometimes consist of the probabilities of two events A and B and of the conditional probability $P(A|B)$. It is then asked that the conditional probability $P(B|A)$

be computed. The corresponding formula is easily obtained from the definition of conditional probability since $P(A \cap B) = P(B|A)P(A) = P(A|B)P(B)$. We therefore have Bayes' retrodiction formula

$$P(B|A) = \frac{P(A|B)P(B)}{P(A)}, \qquad (32)$$

which allowed Bayes to compute the probability of the "cause" B given the "consequence" A (hence the terminology). Of course, in the above equalities we have implicitly assumed that $P(A)$ and $P(B)$ are strictly positive so that we can speak of quantities such as $P(B|A)$ and $P(A|B)$. In practice, the problem of retrodiction arises when A and B are probabilistically compatible, i.e., $P(A \cap B) > 0$, in which case $P(A)$ and $P(B)$ are actually strictly positive.

Bayes' Sequential Formula. Let $A_1, \ldots, A_n$ be events such that $P(A_1 \cap \cdots \cap A_n) > 0$. Then

$$P(A_1, \ldots, A_n) = P(A_1)P(A_2|A_1)P(A_3|A_1, A_2) \cdots P(A_n|A_1, \ldots, A_{n-1}). \qquad (33)$$

Here, notation $P(A_1, \ldots, A_n)$ is equivalent to $P(\bigcap_{k=1}^{n} A_k)$. Similarly, $P(B|A_1, \ldots, A_k) = P(B|\bigcap_{j=1}^{k} A_j)$.

PROOF OF EQ. (33). The proof is by induction: for $n = 2$, formula (33) is just the definition of conditional probability, and if Eq. (33) is true for some $n \geqslant 2$, it is also true for $n + 1$, since

$$P(A_1, \ldots, A_{n+1}) = P\left(\left(\bigcap_{j=1}^{n} A_j\right) \cap A_{n+1}\right) = P\left(\bigcap_{j=1}^{n} A_j\right) P\left(A_{n+1} \middle| \bigcap_{j=1}^{n} A_j\right)$$

$$= P(A_1, \ldots, A_n)P(A_{n+1}|A_1, \ldots, A_n). \qquad \square$$

Formula (33) has an appealing intuitive interpretation if one identifies an index n with a time. Thus $A_1, A_2, A_3, \ldots$ are events happening (or not happening) in sequence, at times $1, 2, 3, \ldots$, respectively.

In the case where at each time $k(2 \leqslant k \leqslant n)$ the conditional probability $P(A_k|A_1, \ldots, A_{k-1})$ is equal to $P(A_k|A_{k-1})$, the sequence $(A_k, 1 \leqslant k \leqslant n)$ is called a *Markovian chain* of events. Equality $P(A_k|A_1, \ldots, A_{k-1}) = P(A_k|A_{k-1})$ expresses the fact that event A_k is conditionally independent of $A_1, \ldots, A_{k-2}$ given A_{k-1} [see Eq. (23)].

The Formula of Incompatible and Exhaustive Causes. Let A be some event, and let $(B_n, n \geqslant 1)$ be an exhaustive sequence of mutually incompatible events. By this we mean that whenever $i \neq j$, $B_i \cap B_j = \varnothing$ (incompatibility) and that $\bigcup_{n=1}^{\infty} B_n = \Omega$ (exhaustivity). In other words, one and only one among the events B_n must happen. We then have the important formula, which we quote together with its condition of application,

$$P(A) = \sum_{n=1}^{\infty} P(A|B_n)P(B_n) \qquad \text{where} \quad \sum_{n=1}^{\infty} B_n = \Omega \quad . \tag{34}$$

To be correct, since $P(A|B_n)$ is defined only when $P(B_n) > 0$, we must agree by convention that in the above formula, $P(A|B_n)P(B_n) = 0$ whenever $P(B_n) = 0$.

PROOF OF EQ. (34). One writes the set identity

$$A = A \cap \Omega = A \cap \left(\sum_{n=1}^{\infty} B_n \right) = \sum_{n=1}^{\infty} A \cap B_n.$$

By the σ-additivity axiom, $P(A) = \sum_{n=1}^{\infty} P(A \cap B_n)$. Now if $P(B_n) = 0$, $P(A \cap B_n) = 0$ since $B_n \subset A \cap B_n$. And if $P(B_n) > 0$, $P(A \cap B_n) = P(A|B_n)P(B_n)$ by definition of the symbol $P(A|B)$. $\qquad \square$

EXAMPLE 8. In a digital communications system, one transmits 0's and 1's through a "noisy" channel that performs as follows: with probability p the transmitted and the received digits are different. It is called a *binary symmetric channel* (see Fig. 10). Suppose that a 0 is emitted with probability π_0 and a 1 with probability $\pi_1 = 1 - \pi_0$. What is the probability of obtaining 1 at the receiving end?

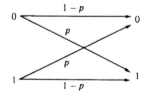

Figure 10. The binary symmetric channel.

Solution. Call X and Y the input and output random variables. Then

$$P(Y = 1) = P(Y = 1|X = 0)P(X = 0) + P(Y = 1|X = 1)P(X = 1)$$

that is

$$P(Y = 1) = p \cdot \pi_0 + (1 - p) \cdot \pi_1.$$

5.2. A Small Bestiary of Exercises

The Intuitive Attack on Probabilistic Problems. One sometimes feels that a complete formalization of a probabilistic problem in the $(\Omega, \mathscr{F}, P)$ framework

is not necessary and that the steps leading to the construction of the probabilistic model can be avoided when a direct formulation in terms of events is
adopted. We will give an instance of this kind of "elementary" problem, which
is usually stated in everyday language.

EXAMPLE 9 (The Bridge Problem). Two locations A and B are linked by three
different paths and each path contains a number of mobile bridges that can
be in the lifted position with a probability indicated in Fig. 11. The bridges
are lifted independently. What is the probability that A is accessible from B,
i.e., that there exists at least one path with no bridge lifted?

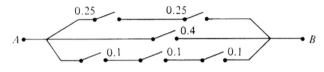

Figure 11. All bridges up.

Solution. The usual (and efficient) approach to this type of problem is to forget
about the formal $(\Omega, \mathscr{F}, P)$ model and directly define the events of interest. For
instance, U_1 will be the event "no bridge lifted in the upper path." Defining
similarly U_2 and U_3, we see that the probability to be computed is that of
$U_1 \cup U_2 \cup U_3$, or by de Morgan's law, that of $\bar{U}_1 \cap \bar{U}_2 \cap \bar{U}_3$, i.e., $1 - P(\bar{U}_1 \cap \bar{U}_2 \cap \bar{U}_3) = 1 - P(\bar{U}_1)P(\bar{U}_2)P(\bar{U}_3)$, where the last equality is obtained in view
of the independence of the bridges in different paths. Letting now $U_1^1 =$
"bridge one in the upper path is not lifted" and $U_1^2 =$ "bridge two in the upper
path is not lifted", we have $U_1 = U_1^1 \cap U_1^2$, therefore, in view of the independence of the bridges, $P(\bar{U}_1) = 1 - P(U_1) = 1 - P(U_1^1)P(U_1^2)$. We must now
use the data $P(U_1^1) = 1 - 0.25$, $P(U_1^2) = 1 - 0.25$ to obtain $P(\bar{U}_1) = 1 - (0.75)^2$. Similarly $P(\bar{U}_2) = 1 - 0.6$ and $P(\bar{U}_3) = 1 - (0.9)^3$. The final result
is $1 - (0.4375)(0.4)(0.271) = 0.952575$.

We now propose a series of exercises stated in nonmathematical language.
The reader will have to interpret the statements and introduce hypotheses of
independence and conditional independence when they are missing and if they
are plausible.

E12 Exercise. To detect veineria (an imaginary disease of the veins), doctors
apply a test, which, if the patient suffers from such disease, gives a positive
result in 99% of the cases. However, it may happen that a healthy subject
obtains a positive result in 2% of the cases. Statistical data show that one
patient out of 1,000 "has it." What is the probability for a patient who scored
positive on the test to be veinerious?

E13 Exercise. Professor Nebulous travels from Los Angeles to Paris with stop
overs in New York and London. At each stop his luggage is transferred from

one plane to another. In each airport, including Los Angeles, chances are that with probability p his luggage is not placed in the right plane. Professor Nebulous finds that his suitcase has not reached Paris. What are the chances that the mishap took place in Los Angeles, New York, and London, respectively?

E14 Exercise. Two factories A and B manufacture watches. Factory A produces on the average one defective item for every 100 items, whereas B produces only one for every 200. A retailer receives a case of watches from one of the two above factories, but he does not know from which one. He checks the first watch and it works. What is the probability that the second watch he will check is good?

E15 Exercise. Two numbers are selected independently at random in the interval $[0, 1]$. You are told that the smaller one is less than $\frac{1}{3}$. What is the probability that the larger one is greater than $\frac{3}{4}$?

E16 Exercise. There are three cards identical in all respects but the color. The first one is red on both sides, the second one is white on both sides, and the third one is red on one side and white on the other (see Fig. 12). A dealer selects one card at random and puts it on the table without looking. Having not watched these operations, you look at the exposed face of the card and see that it is red. What is the probability that the hidden face is also red?

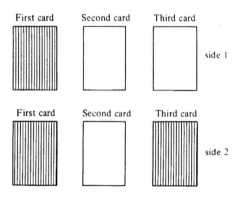

Figure 12. Three cards.

6. Counting and Probability

A number of problems in Probability reduce to counting the elements in a finite set. The general setting is as follows. The set Ω of all possible outcomes is finite, and *all outcomes $\omega \in \Omega$ have the same probability $p(\omega) = P(\{\omega\})$*, which must be equal to $1/\text{card }\Omega^*$ since $\sum_{\omega \in \Omega} p(\omega) = 1$. An event A is a collection of "favorable" outcomes, and its probability is $P(A) = \sum_{\omega \in A} P(\{\omega\})$, so that

$$P(A) = \frac{\text{card } A}{\text{card } \Omega} . \tag{35}$$

Therefore, one must count the elements of A and Ω. The art of counting is called *Combinatorics* and is a rich area of mathematics. We shall now give the first elements of Combinatorics and apply them to simple situations.

Counting Injections (Ordered Arrangements without Repetition). Let E and F be two finite sets, and denote $p = \text{card } E$, $n = \text{card } F$. Also suppose, without

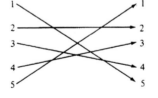

(a) An injection from $E = \{1, 2, 3, 4, 5\}$ into
$F = \{1, 2, 3, 4, 5, 6, 7\}$.

(b) A permutation of $E = \{1, 2, 3, 4, 5\}$.

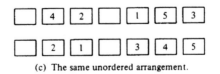

(c) The same unordered arrangement.

Figure 13. Arrangements.

---

* If E is a set, card E is the cardinality of E, that is, when E is finite, the number of elements of E. Another notation is $|E|$.

loss of generality, that $E = \{1, 2, \ldots, p\}$. We are going to count the functions $f: E \to F$, which are *injections*, i.e., such that there exists no pair $i, j \in E$, $i \neq j$, with $f(i) = f(j)$. Clearly, if $p > n$, there is no such function. *If $p \leqslant n$, the number of injections from E into F is*

$$A_n^p = n(n - 1)\ldots(n - p + 1) = \frac{n!}{(n - p)!} .$$
(36)

PROOF. To construct an injection $f: E \to F$, one must first select $f(1)$ in F. There are n possibilities. Now, once $f(1)$ is chosen, there are only $n - 1$ possible choices for $f(2)$ since $f(2)$ must differ from $f(1)$, etc. until we choose $f(p)$ among the $(n - p + 1)$ remaining possibilities, i.e., in $F - \{f(1), \ldots, f(p - 1)\}$. In summary, we have $n(n - 1)\ldots(n - p + 1)$ possibilities for f.   □

Permutations. A special case of interest occurs when $n = p$. In this case, from Eq. (36), we see that $A_n^n = n!$. Now, if card $E =$ card F, an injection of E into F is necessarily a bijection. Recalling that by definition a permutation of E is a bijection of E into itself, we see by specializing to the case $E \equiv F$, that *the number of permutations of a set with n elements is*

$$\boxed{P_n = n!} .$$
(37)

Counting Subsets of a Given Size. Now let F be a finite set with $n =$ card F elements. We ask the question, How many different subsets of p elements $(p \leqslant n)$ are there? If we had asked, How many *ordered* subsets of F with p elements are there?, the answer would have been A_n^p because such an ordered subset

$$\{x_1, \ldots, x_p\} \qquad (x_i \in F)$$

is, since $i \neq j$ implies $x_i \neq x_j$, identifiable with an injection $f: \{1, \ldots, p\} \to F$ defined by $f(i) = x_i$. But for our problem, we have been counting too much. Indeed, all permutations of the ordered subset $\{x_1, \ldots, x_p\}$ represent the same (unordered) subset. Therefore, the number of different subsets of F with p elements is A_p^n divided by the number $p!$ of permutations of a set with p elements.

In summary, *let F be a set with card $F = n$ elements, and let $p \leqslant n$. The number of subsets of F with p elements is*

$$\binom{n}{p} = \frac{n!}{(n - p)!p!} ,$$
(38)

where $\binom{n}{p}$ is a symbol defined by the right-hand side of Eq. (38).

Let now F be a finite set with card $F = n$ elements. How many subsets of

F are there? One could answer with $\sum_{p=0}^{n} \binom{n}{p}$, and this is true if we use the convention that $\binom{n}{0} = 1$ (or equivalently $0! = 1$). [Recall that the void set $\varnothing$ is a subset of F, and it is the only subset of F with 0 elements. Formula (38) is thus valid for $p = 0$ with the above convention.]

We will prove now that the number of subsets of F is 2^n, and therefore

$$2^n = \sum_{p=0}^{n} \binom{n}{p}. \tag{39}$$

Let $x_1, x_2, \ldots, x_n$ be an enumeration of F. To any subset of F there corresponds a sequence of length n of 0's and 1's, where there is a 1 in the ith position if and only if x_i is included in the subset. Conversely, to any sequence of length n of 0's and 1's, there corresponds a subset of F consisting of all x_i's for which the ith digit of the sequence is 1. Therefore, the number of subsets of F is equal to the number of sequences of length n of 0's and 1's, which is 2^n.

The Binomial Formula. Formula (39) is a particular case of the *binomial formula*

$$(x + y)^n = \sum_{p=0}^{n} \binom{n}{p} x^p y^{n-p} \qquad (x, y \in \mathbb{R}). \tag{40}$$

It suffices to let $x = y = 1$ in Eq. (40) to obtain Eq. (39).

PROOF OF EQ. (40). Let x_i, y_i ($1 \leqslant i \leqslant n$) be real numbers. The product $\prod_{i=1}^{n} (x_i + y_i)$ is formed of all possible products $x_{i_1} x_{i_2} \ldots x_{i_p} y_{j_1} \ldots y_{j_{n-p}}$ where $\{i_1, \ldots, i_p\}$ is a subset of $\{1, \ldots, n\}$, and $\{j_1, \ldots, j_{n-p}\}$ is the complement of $\{i_1, \ldots, i_p\}$ in $\{1, \ldots, n\}$. Therefore,

$$\prod_{i=1}^{n} (x_i + y_i) = \sum_{p=0}^{n} \sum_{\substack{\{i_1, \ldots, i_p\} \\ \{i_1, \ldots, i_p\} \subset \{1, \ldots, n\}}} x_{i_1} \ldots x_{i_p} y_{j_1} \ldots y_{j_{n-p}}.$$

The second $\sum$ in the right-hand side of this equality contains $\binom{n}{p}$ elements, since there are $\binom{n}{p}$ different subsets $\{i_1, \ldots, i_p\}$ of p elements of $\{1, \ldots, n\}$. Now letting $x_i = x$, $y_i = y$ ($1 \leqslant i \leqslant n$), we obtain the binomial formula. $\qquad\square$

In view of the symmetric roles of x and y in Eq. (40),

$$\binom{n}{p} = \binom{n}{n-p}. \tag{41}$$

Another important formula is Pascal's formula (Fig. 14):

$$\binom{n}{p} = \binom{n-1}{p-1} + \binom{n-1}{p}. \tag{42}$$

It is obtained by selecting an element $x_0 \in F$ and observing that the subsets of p elements either contain x_0 or do not contain x_0.

n \ p	0	1	2	3	4	5	6	7	8	9	10	...
0	1											
1	1	1										
2	1	2	1									
3	1	3	3	1								
4	1	4	6	4	1							
5	1	5	10	10	5	1						
6	1	6	15	20	15	6	1					
7	1	7	21	35	35	21	7	1				
8	1	8	28	56	70	56	28	8	1			
9	1	9	36	84	126	126	84	36	9	1		
10	1	10	45	120	210	252	210	120	45	10	1	
.	.	.	.	.	.	.	.	.	.	.	.	...

Figure 14. Pascal's array. The entry (n, p) is $\binom{n}{p}$. Pascal's array is constructed as follows: first fill the first column and the diagonal with 1's, and then fill the rest of the lower triangle by applying formula (42).

An "Urn Problem." There is an urn containing N_1 black balls and N_2 red balls. You draw at random n balls from the urn $(n \leqslant N_1 + N_2)$ (Fig. 15). What is the probability that you have k black balls $[0 \leqslant k \leqslant \inf(N_1, n)]$?

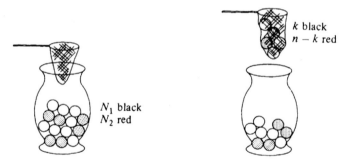

Figure 15. The urn problem.

Solution. The set of outcomes Ω is the family of all subsets ω of n balls among the $N_1 + N_2$ balls in the urn. Therefore,

$$\text{card}\,\Omega = \binom{N_1 + N_2}{n}.$$

Now you must count the subsets ω with k black balls and $n - k$ red balls. To form such a set, you first form a set of k black balls among the N_1 black balls, and there are $\binom{N_1}{k}$ possibilities. To each such subset of k black balls, you must associate a subset of $n - k$ red balls. This multiplies the possibilities by $\binom{N_2}{n-k}$. Thus, if A is the number of subsets of n balls among the $N_1 + N_2$ balls in the urn which consist of k black balls and $n - k$ red balls,

$$\text{card } A = \binom{N_1}{k} \cdot \binom{N_2}{n-k}.$$

The answer is therefore

$$P(A) = \frac{\binom{N_1}{k}\binom{N_2}{n-k}}{\binom{N_1 + N_2}{n}}.$$

E17 Exercise. There are n points on a circumference. Two are chosen randomly (Fig. 16). What is the probability p_n that they are neighbors?

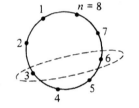

Figure 16. Illustration of Exercise E17. Here $n = 8$, and the pair $(3,6)$ has been drawn.

E18 Exercise. An urn contains N balls numbered from 1 to N. Someone draws n balls $(1 \leqslant n \leqslant N)$ simultaneously from the urn. What is the probability that the lowest number drawn is $k(k \leqslant N - n)$?

7. Concrete Probability Spaces

Some beginners have no problem in accepting the notion of a random variable. For them it is rather intuitive. When speaking of a random variable X, they think: this is just a random number, i.e. "something" that takes random values. And this randomness is somehow embodied in the c.d.f. $F(x) = P(X \leqslant x)$, where $F(b) - F(a)$ measures the "chance" of obtaining a value in the interval $(a, b]$.

Another category of students, with a different psychology, prefer to think of *"something that takes values"* as a function. The $(\Omega, \mathscr{F}, P)$ formalism is made for them, because it says that a random variable X is *just a function*, in the ordinary sense, of $\omega \in \Omega$ [with in addition a rather innocuous requirement, namely (11), but that is not essential at this point of the discussion]. It is not the function X, that is random, but the argument ω; and $X(\omega)$ is the random number, not X.

For the student who prefers to consider X as a function of ω, the nature of

ω must be made precise. If he does not like the vagueness of a phrase like "X is a random number," he will most probably also want to know what $(\Omega, \mathcal{F}, P)$ really is. The discussion to follow is intended to shed some light on this point.

The manner in which Probability was introduced in Section 1 is abstract and axiomatic: one starts with an *abstract probability space* $(\Omega, \mathcal{F}, P)$, which is given but not specified, and where $\mathcal{F}$ and P are assumed to satisfy a few axioms.

In Subsection 2.2 two concrete probability spaces concerning elementary probabilistic models (the three-dice game and the random point in the unit square) were constructed. They are called *concrete probability spaces* because the sample space Ω is described in terms of "concrete" mathematical objects: in the examples of Subsection 2.2., Ω was either the finite set $\{1, 2, 3, 4, 5, 6\}^3$ or the unit square in $\mathbb{R}^2$. Also in Illustration 2 at the end of this chapter, featuring the famous *ballot problem*, Ω is a finite set, a typical element of which is a non-decreasing function $f: \{0, 1, \ldots, b\} \to \{0, 1, \ldots, a\}$ such that $f(b) = a$. In each case, $\mathcal{F}$ and P are *constructed* and the axioms of probability are *verified*.

When the sample space is finite, the construction of $\mathcal{F}$ and P is very simple: take for $\mathcal{F}$ the family of *all* subsets of Ω—and this is a σ-field indeed, associate to each $\omega \in \Omega$ a non-negative number $p(\omega)$ such that $\sum_{\omega \in \Omega} p(\omega) = 1$, and define $P(A) = \sum_{\omega \in A} p(\omega)$. This type of construction is used in the three-dice model and in the ballot problem.

In the example of Subsection 2.2, relative to a random point in the unit square the construction was not too difficult either, although some fine mathematical points have been left aside and taken for intuitively clear. For instance, Ω being a unit square of $\mathbb{R}^2$, namely $[0, 1]^2$, $\mathcal{F}$ was defined to be the family of subsets of Ω for which the area can be defined. Deep and somewhat difficult mathematical results are hidden behind the phrase "for which the area can be defined." The mathematical theory behind it is the Lebesgue Theory of Measure and Integration. It is not in the scope of the present introductory text. As a matter of fact one can profitably study Probability Theory without knowing Integration Theory, at least up to a certain point. Of course knowledge of Integration Theory helps and sometimes it becomes a necessity, but only in the more advanced topics.

Lebesgue theory states, in the particular case of interest to us, that there exists a σ-field $\mathcal{F}$ on $\Omega = [0, 1]^2$ for which the "area" can be defined. The "area" of A is called the *Lebesgue measure* of A. Of course for rectangles and for subsets of A with a familiar shape (triangles, circles, etc.) the Lebesgue measure coincides with the area as it is defined in high school mathematics. So why use Lebesgue theory when high school mathematics suffices? An answer is: Lebesgue theory is able to consider sample spaces much more complex than a square; an example will be given soon. Another answer, more technical, is the following: the class of subsets of the square for which the elementary area can be defined is not a σ-field. This is why Lebesgue defined $\mathcal{F}$ to be the smallest σ-field containing all the rectangles in the square Ω. It

is not difficult to show that such $\mathscr{F}$ exists, but the problem with this definition
is that it is not constructive, and therefore one can define the Lebesgue measure
of a set in $\mathscr{F}$ only in a nonconstructive way. Lebesgue proved the following
theorem: there exists a unique set function $P: \mathscr{F} \rightarrow [0, 1]$ associating with each
$A \in \mathscr{F}$ its Lebesgue measure $P(A)$, a set function that is σ-additive and is such
that if A is a rectangle $[a, b] \times [c, d]$, then $P[A] = (b - a)(c - d)$, the area of
A. Moreover, the Lebesgue measure and the area are the same for "ordinary"
sets such as sets bounded by a piecewise smooth curve.

Now one can wonder: why should one insist on having a σ-field of events?
Who cares to compute the probability that the random point ω falls into a
pathological set that cannot even be described?

In the specific example concerning a random point in the unit square it is
true that nobody really needs to have a σ-field of events. But in the abstract
definition of $(\Omega, \mathscr{F}, P)$, the σ-field property of $\mathscr{F}$ cannot be dispensed with.
This will now be explained in one of the most interesting models of Probability
Theory from the theoretical point of view.

An Infinite Sequence of Heads and Tails Played with a Fair Coin. Here, Ω is
the interval $(0, 1]$, $\mathscr{F}$ is the smallest σ-field on Ω containing all the segments
$[a, b]$ of $(0, 1]$, and for any $A \in \mathscr{F}$, $P(A)$ is the Lebesgue measure ("length") of
A, that is to say the unique probability measure P on $(\Omega, \mathscr{F})$ such that for any
$A = [a, b] \subset (0, 1]$, $P(A) = b - a$. Here again the existence and uniqueness of
such P is a theoretical result of Measure Theory that will be accepted without
proof in this book.

It is claimed that $(\Omega, \mathscr{F}, P)$, so constructed, aptly models not only a random
point on the unit segment but also infinite games of heads and tails with a fair
coin. This claim will now be examined.

Each $\omega \in (0, 1]$ can be expressed in binary form as

$$\omega = 0 \cdot \omega_1 \omega_2 \omega_3 \ldots \tag{43}$$

where ω_n is 0 or 1. Such development is called dyadic, and it is unique if one
requires that there be an infinity of 1's in it for any ω. For instance the number
$\omega = \frac{1}{4}$ will be written not as 0.01000... but instead as 0.00111111.... Fig. 17
shows how ω_n is obtained from ω.

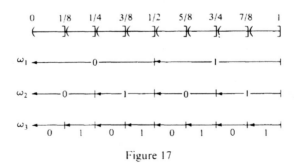

Figure 17

Now define for any n the random variable X_n by

$$X_n(\omega) = \omega_n. \tag{44}$$

That is, $X_n(\omega)$ is the nth term in the dyadic development of $\omega \in (0, 1]$. The set $\{\omega \mid X_n(\omega) = 0\}$ is the sum $\sum_{k=0}^{2^{n-1}-1} (2k \cdot 2^{-n}, (2k + 1)2^{-n}]$ with total length $\frac{1}{2}$. Thus $P(X_n = 0) = \frac{1}{2}$, and similarly $P(X_n = 1) = \frac{1}{2}$.

The event $\{X_1 = 1, X_2 = 0, X_3 = 1\}$ is the subset of $(0, 1]$ pictured in Fig. 18, namely $(\frac{5}{8}, \frac{3}{4}]$.

$$\omega = 0.101 \ XXXX \ldots$$

Figure 18. The event $\{X_1 = 1, X_2 = 0, X_3 = 1\}$.

Its probability (its length) is $\frac{1}{8}$ and therefore $P(X_1 = 1, X_2 = 0, X_3 = 1) = P(X_1 = 1)P(X_2 = 0)P(X_3 = 1)$. More generally, it is seen that $P(X_1 = a_1, X_2 = a_2, \ldots, X_n = a_n) = (\frac{1}{2})^n$ for any sequence $a_1, a_2, \ldots, a_n$ of 0's and 1's. Therefore

$$\boxed{P(X_1 = a_1, \ldots, X_n = a_n) = P(X_1 = a_1) \ldots P(X_n = a_n)}. \tag{45}$$

This shows that for all $n \geqslant 1$ the random variables $X_1, \ldots, X_n$ are *independent*, with values 0 and 1, and that $P(X_n = 1) = P(X_n = 0) = \frac{1}{2}$.

Now if X_n is interpreted as the result of the nth toss of a fair coin (say 0 for "tails" and 1 for "heads"), we have a concrete probabilistic model for the game of heads and tails with a fair coin. The concreteness of the model is, one must admit, very relative, but the interval $(0, 1]$ is a somewhat concrete object for a mathematician. But even a non-mathematician should be satisfied with such a model because it features the nth toss of the coin (the random variable X_n), the coin is fair (Pr. (heads) = Pr. (tails) = $\frac{1}{2}$) and the successive tosses are independent.

Note that once God has selected ω on $(0, 1]$ at random, the whole sequence $X_1(\omega), X_2(\omega), \ldots$ is known to Him, but He will show it to you progressively, toss after toss. Any other probabilistic model for an infinite game of heads and tails must feature random variables X_n which are functions of ω, and therefore ω is first drawn from Ω *and then* the values $X_1(\omega), X_2(\omega), \ldots$ —the values of the functions $X_1, X_2, \ldots$ at ω—are "instantly" available, although in practice index n plays the role of time and $X_n(\omega)$ is shown to you only at the nth stage of the game.

It must now be checked that this probability model is in accord with our intuition of probability as idealization of empirical frequency. More explicitly consider the random variable Z_n defined by $Z_n(\omega) = (X_1(\omega) + \cdots + X_n(\omega))/n$. This is the *empirical frequency* of heads in n tosses.

It is "known from experience" that $Z_n(\omega)$ tends to $\frac{1}{2}$ as n goes to ∞. Probabilists add "with probability 1." They do this in order to take care of an ω such as $\omega = 0.001111111111\ldots$ that is to say $\omega = \frac{1}{4}$, and of other ω's of a pathological kind for which $Z_n(\omega)$ does not tend to $\frac{1}{2}$ as $n \to \infty$. The claim of Probabilists is that such ω's are indeed pathological in the sense that

$$\boxed{P(\{\omega | \lim Z_n(\omega) = \tfrac{1}{2}\}) = 1} \qquad (46)$$

This is the famous *strong law of large numbers* for heads and tails. It is a physical law in some sense, but here, in the mathematical setting of Probability Theory, it becomes a *theorem*, and it was proved by Borel in 1909. The proof is given in Chapter 5.

Now, look at the subset of $(0, 1]$, the set of ω's such that $\lim Z_n(\omega) = \frac{1}{2}$. It is not a set for which one can define a length as one would do in high school. Borel's strong law of large numbers states that its Lebesgue measure (length) is 1, the length of the whole segment $(0, 1]$. But yet it is not the whole segment $(0, 1]$. A lot of ω's in $(0, 1]$ are such that $\lim Z_n(\omega)$ does not exist or, if it exists, is not equal to $\frac{1}{2}$. For instance any ω of the form $\omega = k/2^n$ is such that $\lim Z_n(\omega) = 1 \neq \frac{1}{2}$, and there are many such ω's, so many that in a segment of arbitrarily small length there is an infinity of them. This should convince you that high school length is out, at least for our purpose, because in elementary length theory, sets with so many holes are not considered.

Also the set $\{\omega | \lim Z_n(\omega) = \frac{1}{2}\}$ is not obtainable by application of a finite number of elementary set operations ($\cup$, $\cap$ and complementation) to intervals of $\Omega = (0, 1]$, but it can nevertheless be shown to be in $\mathscr{F}$, the smallest σ-field on Ω containing the intervals. It is for that kind of reason that one wants to consider abstract probability spaces $(\Omega, \mathscr{F}, P)$ for which $\mathscr{F}$ is a σ-field. It is a natural structure, especially when one has to consider infinite sequences of random variables, as will be done in Chapter 5, where the convergence of such sequences is studied.

The concrete probability space $(\Omega, \mathscr{F}, P)$ where $\Omega = (0, 1]$ and P is the "length" or Lebesgue measure is just one among the probabilistic models of an infinite game of heads and tails with a fair coin. There are many other models available for this game. One of them will now be briefly described.

Take Ω to be the set of sequences taking their values in $\{0, 1\}$: $\omega = (\omega_n, n \geq 1)$ where $\omega_n = 0$ or 1. Define X_n by $X_n(\omega) = \omega_n$ and let $\mathscr{F}$ be the smallest σ-field on Ω that contains the subsets $\{\omega | \omega_n = 1\}$, $n \geq 1$. Since it is a σ-field, it also contains the set $\{\omega | \omega_n = 0\}$ the complement of $\{\omega | \omega_n = 1\}$, and the sets $\{\omega | \omega_1 = a_1, \ldots, \omega_n = a_n\}$ for all $n \geq 1$ and any sequence $a_1, \ldots, a_n$ taking the values 0 or 1.

Since $\{\omega | \omega_n = a_n\} = \{\omega | X_n(\omega) = a_n\}$, we see that when Ω is equipped with the σ-field $\mathscr{F}$, the X_n's become random variables.

Now a theorem of Measure Theory will be invoked: it says that there exists one and only one probability measure P on $(\Omega, \mathscr{F})$ such that

Illustration 1. A Simple Model in Genetics   31

$$P(\{\omega \mid X_1(\omega) = a_1, \ldots, X_n(\omega) = a_n\}) = (\tfrac{1}{2})^n.$$ (47)

for all $n \geqslant 1$, all $a_1, \ldots, a_n \in \{0, 1\}$.

Thus we have obtained another model for the infinite game of heads and tails with a fair coin. This model is more natural than the previous one because here $\omega = (X_1(\omega), X_2(\omega), \ldots)$ i.e. ω *is the result of the infinite sequence of tosses.* However, we have been forced to take for granted a difficult result of Measure Theory, namely the existence and uniqueness of P satisfying (47). In the concrete model where $\Omega = (0, 1]$ we have also taken for granted the existence of the Lebesgue measure ("length"). The fact is that in virtually every concrete probability space where Ω is not finite or denumerable, a theorem of existence must be invoked. Measure theory provides such existence theorems, and some of them are so powerful that one should not worry about the existence of a concrete probability space, at least in the probabilistic models that will be encountered in this book.

Illustration 1.   A Simple Model in Genetics: Mendel's Law and Hardy–Weinberg's Theorem

In diploid organisms (you are one of them!) each hereditary character is carried by a pair of genes. We will consider the situation in which each gene can take two forms called alleles, denoted a and A. Such was the case in the historical experiments performed in 1865 by the Czech monk Gregory Mendel who studied the hereditary transmission of the nature of the skin in a species of green peas. The two alleles corresponding to the gene or character "nature of the skin" are a for "wrinkled" and A for "smooth". The genes are grouped into pairs and there are two alleles, thus three genotypes are possible for the character under study: aa, Aa (same as aA), and AA. With each genotype is associated a phenotype which is the external appearance corresponding to the genotype. Genotypes aa and AA have different phenotypes (otherwise no character could be isolated), and the phenotype of Aa lies somewhere between the phenotypes of aa and AA. Sometimes, an allele is dominant, e.g., A, and the phenotype of Aa is then the same as the phenotype of AA.

During the reproduction process, each of the two parents contributes to the genetic heritage of their descendant by providing *one* allele of their pair. This is done by the intermediary of the reproductive cells called gametes (in the human species, the spermatozoid and the ovula) which carry only one gene of the pair of genes characteristic of each parent. The gene carried by the gamete is chosen at random among the pair of genes of the parent. The selection procedure for the genotype of the descendant is summarized in Fig. 19. The actual process occurring in the reproduction of diploid cells is called *meiosis*.

| Parent 1
has genotype
Aa | Parent 2
has genotype
AA |

| Gamete 1
carries allele
a
(it had the "choice"
between *A* and *a*) | Gamete 2
carries allele
A
(it had no
other choice) |

Gametes 1 and 2 unite
and provide
descendant with
genotype
Aa

Figure 19. The selection of a genotype.

A given cell possesses two chromosomes. A chromosome can be viewed as a string of genes, each gene being at a specific location in the chain (Fig. 20).

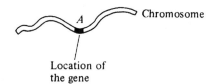

Figure 20. A schematic representation of a chromosome.

The chromosomes double and four new cells are formed for every chromosome (Fig. 21).

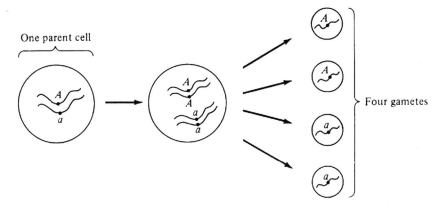

Figure 21. Meiosis.

Illustration 2. The Art of Counting   33

Let us start from an idealistically infinite population where the genotypes are found in the following proportions

$$AA \quad Aa \quad aa$$
$$x \; : \; 2z \; : \; y \; .$$

Here x, y, and z are numbers between 0 and 1, and

$$x + 2z + y = 1.$$

The two parents are chosen independently (random mating), and their gamete chooses an allele at random in the pair carried by the corresponding parent.

E19 Exercise. What is the genotypic distribution of the second generation? Numerical applications: $x = y = 2z = \frac{1}{3}$; $x = y = z = \frac{1}{4}$; $x = \frac{1}{4}$, $y = \frac{1}{2}$, $2z = \frac{1}{4}$; $x = y = \frac{1}{2}$, $z = 0$.

E20 Exercise. Show that the genotypic distributions of all generations, starting from the third one, are the same. (This result was discovered by Hardy and Weinberg.) Show that the stationary distribution depends only on the proportion c of alleles of type A in the initial population.

Illustration 2.   The Art of Counting: The Ballot Problem and the Reflection Principle

In an election, candidates I and II have obtained a and b votes respectively. Candidate I won, that is, $a > b$. What is the probability that in the course of the vote counting procedure, candidate I has always had the lead?

Solution. The vote counting procedure is represented by a path from $(0,0)$ to (b,a) (Fig. 22). Therefore, we shall identify an outcome ω of the vote counting procedure to such a path. The set of all possible outcomes being Ω, we shall prove later that

$$\text{card } \Omega = \binom{a+b}{a} = \binom{a+b}{b}. \qquad (48)$$

Let A be the set of paths of Ω that do not meet the diagonal. This represents the set of favorable outcomes, i.e., the outcomes for which A has the lead throughout the vote-counting procedure. The path ω of Fig. 22(i) is not a favorable path, whereas that of Fig. 22(ii) is a favorable path.

We must now evaluate $\text{card } A$ to find $P(A)$ according to Eq. (35). The following trick is proposed. We consider three disjoint subsets of Ω (Fig. 23):

A has already been defined.

B is the set of unfavorable paths that start well for candidate I, i.e., the first ballot out of the box bears the name of I.

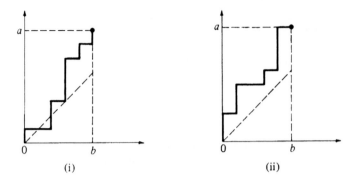

Figure 22. A schematic representation of the vote counting procedure (in (ii), candidate I leads throughout).

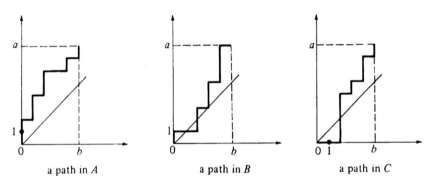

Figure 23. In cases B and C, I does not lead throughout. In case B the first vote is favorable to I, whereas in case C it is favorable to II.

C is the set of unfavorable paths that start well for candidate II.

Clearly,

$$\text{card } \Omega = \text{card } A + \text{card } B + \text{card } C. \tag{49}$$

If we admit Eq. (48), card C is easy to evaluate. It is the number of paths from $(1, 0)$ to (b, a) or, equivalently, the number of paths from $(0, 0)$ to $(b - 1, a)$, i.e.,

$$\text{card } C = \binom{a + b - 1}{b - 1} = \binom{a + b - 1}{a}. \tag{50}$$

It turns out (and we shall prove it below) that

$$\text{card } B = \text{card } C. \tag{51}$$

Therefore, in view of Eqs. (49) and (51),

Illustration 2. The Art of Counting   35

$$\frac{\text{card } A}{\text{card } \Omega} = 1 - 2\frac{\text{card } C}{\text{card } \Omega},$$

and in view of Eqs. (48) and (50),

$$P(A) = \frac{\text{card } A}{\text{card } \Omega} = 1 - 2\frac{\binom{a + b - 1}{a}}{\binom{a + b}{a}} = 1 - 2\frac{(a + b - 1)!a!b!}{(a + b)!a!(b - 1)!}.$$

Finally,

$$P(A) = \frac{a - b}{a + b}. \tag{52}$$

The proof will be complete if we show that Eqs. (48) and (51) are indeed true.

PROOF OF EQ. (51) (The "Reflection Principle"). The proof consists in finding a bijection between B and C since a bijection between two sets exist if and only if the two sets have the same cardinality. The bijection $f : B \to C$ is defined as shown in Fig. 24. A path $\omega \in B$ must meet the diagonal. Let (u, u) be the first point at which $\omega \in B$ meets the diagonal. Then $f(\omega)$ and ω are the same from (u, u) to (b, a), and $f(\omega)$ and ω are symmetric with respect to the diagonal from $(0, 0)$ to (u, u). The mapping f is clearly onto and into, i.e., bijective. $\square$

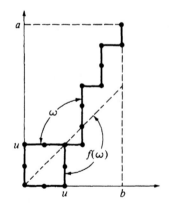

Figure 24. A path and its reflection.

PROOF OF EQ. (48). A vote-counting procedure is representable by a sequence of length $a + b$ of I's and II's, with a I's and b II's. The interpretation of such a sequence is that you find I in position i if and only if the ith ballot is in favor of I. The positions occupied by the I's in a given such sequence is $\{i_1, \ldots, i_a\} \subset \{1, \ldots, a + b\}$. Hence, Eq. (48) in view of Eq. (38).

Illustration 3. Bertrand's Paradox

A random chord CD is drawn on the unit circle. What is the probability that
its length exceeds $\sqrt{3}$, the length of the side of the equilateral triangle inscribed
in the unit circle (see Fig. 25)?

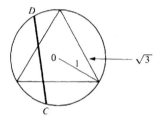

Figure 25. The random chord.

This is a famous example of an incompletely specified problem. One must
tell what "random chord" means. How is it "randomly"chosen? The French
mathematician Bertrand proposed three different answers, and this is the so-
called Bertrand's paradox. The different answers correspond to actually different probabilistic models, i.e different concrete probability spaces $(\Omega, \mathcal{F}, P)$.

First Model. Take for Ω the unit disk with the σ-field $\mathcal{F}$ of "subsets for which
area is defined," and let for any $A \in \mathcal{F}$, $P(A) =$ area of A divided by the area
of the unit disk. This is indeed a probability since $P(\Omega) = 1$, and P is a multiple
of the Lebesgue measure (see Section 7) on the disk. Now C and D are
constructed as Fig. 26 indicates, i.e CD is perpendicular to 0ω. The length of
CD is called $X(\omega)$ since it is a function of ω. This defines a random variable,
and we want to compute $P(X \geqslant \sqrt{3})$. But the event $\{\omega | 0\omega \geqslant \frac{1}{2}\}$ is the shaded
domain of Fig. 26(b). Thus $P(X \geqslant \sqrt{3}) = \frac{3}{4}$. Therefore (first answer) the probability asked by Bertrand is $\frac{3}{4}$.

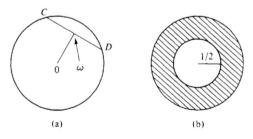

(a)   (b)

Figure 26. First construction of a random chord.

Second Model. Take for Ω the unit circle with the σ-field $\mathcal{F}$ of "subsets for
which length can be defined," and for any $A \in \mathcal{F}$, $P(A) =$ length of A divided

Illustration 3. Bertrand's Paradox 37

by 2π (note that $P(\Omega) = 1$ as required). The random sample ω is a point of the circle. C being *fixed*, take $D = \omega$ (see Fig. 27). Thus CD is indeed a random chord (depending on ω) with a random length $X(\omega)$. The set $\{\omega | X(\omega) \geqslant \sqrt{3}\}$ is the portion of the unit circle enhanced in Fig. 27(b), and the second answer to Bertrand's problem is therefore $\frac{1}{3}$.

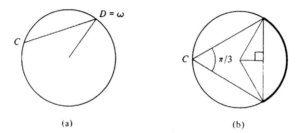

Figure 27. Second construction of a random chord.

Third Model. Take $\Omega = [0, 1]$ (see Figure 28) with the σ-field $\mathscr{F}$ of "subsets for which the length can be defined," and let $P(A) = $ length of A. Define CD to be the chord passing through ω and perpendicular to the Ox axis. It is clear that the length $X(\omega)$ of CD exceeds $\sqrt{3}$ if and only if $\omega \in [\frac{1}{2}, 1]$. Thus the third answer to Bertrand's problem is $\frac{1}{2}$.

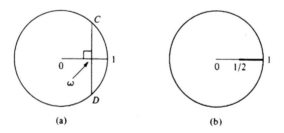

Figure 28. Third construction of a random chord.

Thus we have obtained 3 answers: $\frac{3}{4}$, $\frac{1}{3}$, and $\frac{1}{2}$! There is however nothing really surprising about this, since the concrete probability models corresponding to the above answers are different. Which one is the "good" one is another question. The correct model depends on the device used to throw a chord at random. The three devices used above are purely intellectual, and most likely, do not correspond to any physical device. In order to discriminate between competing probabilistic models one must resort to statistical analysis which is essentially based on two results of Probability Theory: the strong law of large numbers and the central limit theorem. This will be discussed i Chapter 5.

SOLUTIONS FOR CHAPTER 1

E1. From the σ-additivity axiom, $P(\bigcup_{n=1}^{\infty} A_n) = P(\sum_{n=1}^{\infty} A_n') = \sum_{n=1}^{\infty} P(A_n')$. Also $A_n' \subset A_n$, and therefore by the monotonicity property (5), $P(A_n') \leqslant P(A_n)$.

E2. We do the second formula only. If we write $P(A) + P(B) + P(C)$ and compare it with $P(A \cup B \cup C)$, we see that we have counted the striped area twice and the dotted area three times. The dotted area is $P(A \cap B \cap C)$, and the striped area is $P(A \cap B) + P(A \cap C) + P(B \cap C)$ minus three times the dotted area. Finally, $P(A \cup B \cup C) = P(A) + P(B) + P(C) - 2P(A \cap B \cap C) - (P(A \cap B) + P(A \cap C) + P(B \cap C) - 3P(A \cap B \cap C))$, which gives the desired result after simplification. *Comment:* Try formalizing the above proof.

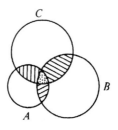

E3. If $a \geqslant 2$, $\{Z \leqslant a\} = \Omega$. If $a < 0$, $\{Z \leqslant a\} = \varnothing$. If $a \in [0, 2]$:

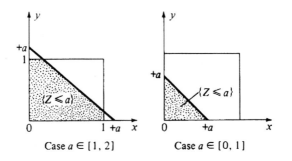

Case $a \in [1, 2]$ Case $a \in [0, 1]$

This figure shows that the set $\{Z \leqslant a\}$ is indeed a set for which the area is well defined.

$$a \geqslant 2, \qquad P(\{Z \leqslant a\}) = P(\Omega) = 1$$

$$a < 0, \qquad P(\{Z \leqslant a\}) = P(\varnothing) = 0$$

$$a \in [0, 1], \qquad P(\{Z \leqslant a\}) = \frac{a^2}{2}.$$

$$a \in [1, 2], \qquad P(\{Z \leqslant a\}) = 1 - \frac{(2 - a)^2}{2}.$$

E4.

$$F(x) = \begin{cases} 0 & \text{if } x < 0 \\ \dfrac{x^2}{2} & \text{if } x \in [0,1] \\ 1 - \dfrac{(2-x)^2}{2} & \text{if } x \in [1,2] \\ 1 & \text{if } x \geq 2 \end{cases}$$

$$f(x) = \frac{dF(x)}{dx} = \begin{cases} 0 & \text{if } x < 0 \\ x & \text{if } x \in [0,1] \\ 2-x & \text{if } x \in [1,2] \\ 0 & \text{if } x \geq 2. \end{cases}$$

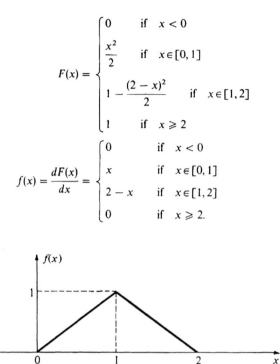

E5. The only way to obtain $X = 18$ is $X_1 = X_2 = X_3 = 6$. The corresponding ω is $\underline{\omega} = \{6, 6, 6,\}$. Therefore,

$$P(X = 18) = P(\{\omega | X(\omega) = 18\}) = P(\{\underline{\omega}\}) = (\tfrac{1}{6})^3.$$

There are 10 different ways of obtaining $X = 6$:

$$\begin{array}{llll} X_1 = 1, & X_2 = 1, & X_3 = 4 & \omega_1 = (1,1,4) \\ X_1 = 1, & X_2 = 4, & X_3 = 1 & \omega_2 = (1,4,1) \\ X_1 = 4, & X_2 = 1, & X_3 = 1 & \omega_3 = (4,1,1) \\ X_1 = 2, & X_2 = 1, & X_3 = 3 & \omega_4 = (2,1,3) \\ X_1 = 1, & X_2 = 2, & X_3 = 3 & \omega_5 = (1,2,3) \\ X_1 = 2, & X_2 = 3, & X_3 = 1 & \omega_6 = (2,3,1) \\ X_1 = 3, & X_2 = 1, & X_3 = 2 & \omega_7 = (3,1,2) \\ X_1 = 1, & X_2 = 3, & X_3 = 2 & \omega_8 = (1,3,2) \\ X_1 = 3, & X_2 = 2, & X_3 = 1 & \omega_9 = (3,2,1) \\ X_1 = 2, & X_2 = 2, & X_3 = 2 & \omega_{10} = (2,2,2) \end{array}$$

Therefore, $P(X = 6) = P(\{\omega | X(\omega) = 6\}) = P(\{\omega_1, \omega_2, \ldots, \omega_{10}\}) = \sum_{j=1}^{10} P(\{\omega_j\}) = 10 \cdot (\tfrac{1}{6})^3$.

E6. Apply the inclusion exclusion formula (see E2) to obtain $P(A \cap B)$ and compare it to $P(A)P(B)$.

	$P(A \cap B)$	$P(A)P(B)$	Independence
Case 1	0.09	0.09	Yes
Case 2	0.24	0.24	Yes
Case 3	0.07	0.15	No

E7. We must check, for instance, that $P(\bar{A}_1 \cap A_2 \cap A_3) = P(\bar{A}_1)P(A_2)P(A_3)$. But $P(A_1 \cap A_2 \cap A_3) = P(A_2 \cap A_3) - P(A_1 \cap A_2 \cap A_3)$ since $\bar{A}_1 \cap A_2 \cap A_3 = A_2 \cap A_3 - A_1 \cap A_2 \cap A_3$. Therefore, $P(\bar{A}_1 \cap A_2 \cap A_3) = P(A_2)P(A_3) - P(A_1)P(A_2)P(A_3) = (1 - P(A_1))P(A_2)P(A_3) = P(\bar{A}_1)P(A_2)P(A_3)$.

E8. $P(A) = P(\{\omega_1, \omega_2\}) = P(\{\omega_1\} + \{\omega_2\}) = \frac{1}{4} + \frac{1}{4} = \frac{1}{2}$. Similarly, $P(B) = P(C) = \frac{1}{2}$. Therefore, $P(A \cap B \cap C) = P(\emptyset) = 0 \neq P(A)P(B)P(C)$. However, $P(A \cap B) = P(\{\omega_2\}) = \frac{1}{4} = P(A)P(B)$ and similarly for $A \cap C$ and $B \cap C$.

E9. Clearly $P_C(A) \geq 0$. Also

$$P_C(A) = \frac{P(A \cap C)}{P(C)} \leq 1$$

since $P(A \cap C) \leq P(A)$.

$$P_C(\Omega) = \frac{P(\Omega \cap C)}{P(C)} = \frac{P(C)}{P(C)} = 1.$$

Let $\{A_n, n \geq 1\}$ be a sequence of disjoint events. Then

$$P_C\left(\sum_{n=1}^{\infty} A_n\right) = \frac{P\left(\left(\sum_{n=1}^{\infty} A_n\right) \cap C\right)}{P(C)} = \frac{P\left(\sum_{n=1}^{\infty} (A_n \cap C)\right)}{P(C)}$$

$$= \sum_{n=1}^{\infty} \frac{P(A_n \cap C)}{P(C)} = \sum_{n=1}^{\infty} P_C(A_n).$$

E10. $$P_C(A|D) = \frac{P_C(A, D)}{P_C(D)} = \frac{P(A, D|C)}{P(D|C)}$$

$$= \frac{P(A, C, D)/P(C)}{P(C, D)/P(C)} = \frac{P(A, C, D)}{P(C, D)} = P(A|C, D).$$

E11. One way of obtaining $S_n = k$ is $X_{i_1} = 1, \ldots, X_{i_k} = 1$ where $1 \leq i_1 < i_2 < \cdots < i_k \leq n$ and $X_j = 0$ for $j \neq i_1, i_2, \ldots, i_k$. The probability of such an event is, by the independence assumption, $P(X_{i_1} = 1) \ldots P(X_{i_k} = 1)P(X_{j_1} = 0) \ldots P(X_{j_{n-k}} = 0)$ [where $\{j_1, \ldots, j_{n-k}\}$ is the set $\{1, 2, \ldots, n\} - \{i_1, \ldots, i_k\}$] i.e., $p^k(1 - p)^{n-k}$. Now the event $S_n = k$ is the sum of all the above events $\{X_{i_1} = 1, \ldots, X_{i_k} = 1, X_{j_1} = 0, \ldots, X_{j_{n-k}} = 0\}$ for all sets of indices $\{i_1, \ldots, i_k\}$ such that $1 \leq i_1 < i_2 < \cdots <$

$i_k \leqslant n$. There are $n!/[k!(n-k)!]$ such sets of indices. Therefore. by the σ-additivity axiom

$$P(S_n = k) = \sum_{1 \leqslant i_1 < \cdots < i_k \leqslant n} P(X_{i_1} = 1, \ldots, X_{i_k} = 1, X_{j_1} = 0, \ldots, X_{j_{n-k}} = 0)$$

$$= \sum_{1 \leqslant i_1 < \cdots < i_k \leqslant n} p^k(1-p)^{n-k} = \frac{n!}{k!(n-k)!} p^k(1-p)^{n-k}.$$

E12. Let M be the event "patient is ill," and $+$ and $-$ be the events "test is positive" and "test is negative," respectively. We have the data

$$P(M) = 0.001, \qquad P(+|M) = 0.99, \qquad P(+|\overline{M}) = 0.02,$$

and we must compute $P(M|+)$. By the retrodiction formula (32),

$$P(M|+) = \frac{P(+|M)P(M)}{P(+)}.$$

By the formula of incompatible and exhaustive causes [Eq. (34)],

$$P(+) = P(+|M)P(M) + P(+|\overline{M})P(\overline{M}).$$

Therfore,

$$P(M|+) = \frac{(0.99)(0.001)}{(0.99)(0.001) + (0.02)(0.999)} \simeq \frac{1}{20}.$$

Comment: This is a low probability indeed. The important thing is not to miss a case. In this respect, the test should have a high value for $P(+|M)$ (here 0.99, but for 0.99999 we would still have $P(M|+) \simeq \frac{1}{20}$). The test is probably inexpensive and perhaps this is why such a large $P(+|\overline{M})$ (here 0.02) is accepted. In medical practice, if a patient has a positive test, he or she is subjected to another test with a smaller $P(+|\overline{M})$. The second test will probably be much more expensive than the first one, otherwise it would have been used in the first place. Using the expensive test only as a second test "to be sure" is cost-effective because only a few people obtain a positive result on the first test. Indeed, with the data in the statement, $P(+) = 0.99 \times 0.001 + 0.02 \times 0.99 \simeq 0.021$. You see that it is the quantity $P(+|\overline{M})$ which is crucial in the computation of $P(M|+)$. For instance, if we take 0.002 instead of 0.02, we obtain $P(M|+) \simeq 0.33$, and with 0.0002, we obtain $P(M|+) \simeq 99/119$.

E13. Think of the misplacement procedure as follows: a demoniac probabilist throws three coins independently and denotes 1 for heads and 0 for tails. This results in three random variables X_1, X_2 and X_3, with values in $\{0, 1\}$, and with $P(X_1 = 1) = P(X_2 = 1) = P(X_3 = 1) = p$. If $X_1 = 1$, the misplacement happened in Los Angeles. If $X_1 = 0$ and $X_2 = 1$, it happened in New York, and if $X_1 = 0$ and $X_2 = 0$ and $X_3 = 1$, it happened in London. The event $M =$ "the luggage has been misplaced" is the sum of these three disjoint (incompatible) events and its probability is therefore $P(M) = P(X_1 = 1) + P(X_1 = 0, X_2 = 1) + P(X_1 = 0, X_2 = 0, X_3 = 1)$. It is natural to assume that the staff in different airports misbehave independently of one another, so that $P(M) = P(X_1 = 1) + P(X_1 = 0)P(X_2 = 1) + P(X_1 = 0)P(X_2 = 0)P(X_3 = 1) = p + (1-p)p + (1-p)^2 p = 1 - (1-p)^3$. This result could have been obtained more simply: $P(M) = 1 - P(\overline{M}) = 1 - P(X_1 = 0, X_2 = 0, X_3 = 0) = 1 - P(X_1 = 0)P(X_2 =$

$0) P(X_3 = 0) = 1 - (1 - p)^3$). We want to compute the probabilities x, y, and z for the luggage to be stranded in Los Angeles, New York, and London, respectively, knowing that it does not reach Paris: $x = P(X_1 = 1 | M)$, $y = P(X_1 = 0, X_2 = 1 | M)$, $z = P(X_1 = 0, X_2 = 0, X_3 = 1 | M)$. One finds

$$x = P(X_1 = 1, M)/P(M) = P(X_1 = 1)/P(M) = \frac{p}{1 - (1 - p)^3}$$

$$y = P(X_1 = 0, X_2 = 1, M)/P(M) = P(X_1 = 0, X_2 = 1)/P(M) = \frac{p(1 - p)}{1 - (1 - p)^3}$$

$$z = P(X_1 = 0, X_2 = 0, X_3 = 1, M)/P(M) = P(X_1 = 0, X_2 = 0, X_3 = 1)/P(M)$$

$$= \frac{p(1 - p)^2}{1 - (1 - p)^3}.$$

E14. Let X_n be the state of the nth watch in the case, with $X_n = 1$ if it works and $X_n = 0$ if it does not. Let Y be the factory of origin. We express our a priori ignorance as to where the case comes from by

$$P(Y = A) = P(Y = B) = \tfrac{1}{2}.$$

Also, we assume that given $Y = A$ (respectively, $Y = B$), the states of the successive watches are independent. For instance,

$$P(X_1 = 1, X_2 = 0 | Y = A) = P(X_1 = 1 | Y = A)P(X_2 = 0 | Y = A).$$

We have the data

$$P(X_n = 0 | Y = A) = 0.01, \qquad P(X_n = 0 | Y = B) = 0.005.$$

We are required to compute $P(X_2 = 1 | X_1 = 1)$, that is, $P(X_1 = 1, X_2 = 1)/P(X_1 = 1)$. By the formula of exclusive and exhaustive causes, $P(X_1 = 1, X_2 = 1) = P(X_1 = 1, X_2 = 1 | Y = A)P(Y = A) + P(X_1 = 1, X_2 = 1 | Y = B)P(Y = B) = \tfrac{1}{2}(99/100)^2 + \tfrac{1}{2}(199/200)^2$, and $P(X_1 = 1) = P(X_1 = 1 | Y = A)P(Y = A) + P(X_1 = 1 | Y = B)P(Y = B) = \tfrac{1}{2}(99/100) + \tfrac{1}{2}(199/200)$. Therefore,

$$P(X_2 = 1 | X_1 = 1) = \frac{\left(\dfrac{99}{100}\right)^2 + \left(\dfrac{199}{200}\right)^2}{\dfrac{99}{100} + \dfrac{199}{200}}.$$

Comment: We see that the states of two successive watches are not independent, otherwise $P(X_2 = 1 | X_1 = 1) = P(X_2 = 1) = \tfrac{1}{2}(99/100) + \tfrac{1}{2}(199/200)$. However, the states of two successive watches were supposed to be *conditionnally independent given the factory of origin*.

E15. We can take the model of drawing a point at random in the unit square, since then the random variables taking their values in $[0, 1]$, X and Y, are independent (Example 7) and uniformly distributed (Example 6). We have to compute $P(\sup(X, Y) \geqslant \tfrac{3}{4} | \inf(X, Y) \leqslant \tfrac{1}{3})$, that is, $P(\sup(X, Y) \geqslant \tfrac{3}{4}, \inf(X, Y) \geqslant \tfrac{1}{3})/P(\inf(X, Y) \leqslant \tfrac{1}{3})$. The sets $\{\sup(X, Y) \geqslant \tfrac{3}{4}, \inf(X, Y) \leqslant \tfrac{1}{3}\}$ and $\{\inf(X, Y) \leqslant \tfrac{1}{3}\}$ are pictured in the figure below and have the probabilities (areas) $1/6$ and $5/9$, respectively. Hence, the result $(1/6)/(5/9) = 3/10$.

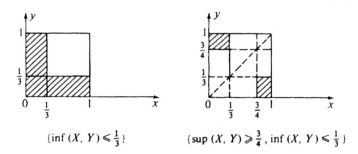

$$\{\inf (X, Y) \leqslant \tfrac{1}{3}\} \qquad\qquad \{\sup (X, Y) \geqslant \tfrac{3}{4}, \inf (X, Y) \leqslant \tfrac{1}{3}\}$$

E16. Call the cards RR, WW, RW. The experiment features two random variables: X is the card selected at random and Y is the color of the exposed face. We must compute $P(X = RR | Y = R)$. The reasonable data consists of $P(X = RR) = P(X = WR) = P(X = WW) = \frac{1}{3}$ and $P(Y = R | X = RR) = 1$, $P(Y = R | X = RW) = \frac{1}{2}$, $P(Y = R | X = WW) = 0$. Now,

$$P(X = RR | Y = R) = P(X = RR, Y = R)/P(Y = R) = P(X = RR)/(P(Y = R)$$

and

$$P(Y = R) = P(Y = R | X = RR)P(X = RR) + P(Y = R | X = WW)P(X = WW)$$
$$+ P(Y = R | X = WR)P(X = WR)$$
$$= 1 \cdot \tfrac{1}{3} + 0 \cdot \tfrac{1}{3} + \tfrac{1}{2} \cdot \tfrac{1}{3} = \tfrac{3}{6} = \tfrac{1}{2}.$$

Therefore, $P(X = RR | Y = R) = \frac{1}{3} : \frac{1}{2} = \frac{2}{3}$ (not $\frac{1}{2}$ as some people guess).

E17. For $n = 2$, $p_2 = 1$. For $n \geqslant 3$, there are n pairs of points which are neighbors. Also, in general, there are $\binom{n}{2}$ pairs. The probability to be found is therefore $n/\binom{n}{2} = 2/n - 1$.

E18. There are $\binom{N}{n}$ subsets of n balls among N balls. If ball k is in the subset and if it is the ball with the lowest number, the remaining $n - 1$ balls must be chosen among $N - k$ balls (i.e., $k + 1, \ldots, N$). This leaves $\binom{N-k}{n-1}$ choices. The probability to be found is therefore $\binom{N-k}{n-1}/\binom{N}{n}$.

E19. The first task consists in providing a probabilistic model. We propose the following one. The sample space Ω is the collection of all quadruples $\omega = (x_1, x_2, y_1, y_2)$ where x_1 and x_2 take their values in $\{AA, aA, aa\}$, and y_1 and y_2 take their values in $\{A, a\}$. The four coordinate random variables X_1, X_2, Y_1, Y_2 are defined by $X_1(\omega) = x_1$, $X_2(\omega) = x_2$, $Y_1(\omega) = y_1$, and $Y_2(\omega) = y_2$. We interpret X_1 and X_2 as the pairs of genes in parents 1 and 2 respectively. Y_1 is the allele chosen by gamete 1 among the alleles of X_1, with a similar definition for Y_2. The data available are

$$\left. \begin{array}{l} P(X_1 = AA) = P(X_2 = AA) = x \\ P(X_1 = aa) = P(X_2 = aa) = y \\ P(X_1 = Aa) = P(X_2 = Aa) = 2z \end{array} \right\} \text{ choice of parents}$$

$$P(Y_1 = A|X_1 = AA) = 1, P(Y_1 = a|X_1 = AA) = 0$$
$$P(Y_1 = A|X_1 = aa) = 0, P(Y_1 = a|X_1 = aa) = 1$$
$$P(Y_1 = A|X_1 = Aa) = \tfrac{1}{2}, P(Y_1 = a|X_1 = Aa) = \tfrac{1}{2}$$

choice of allele by gamete 1

and the similar data for the choice of allele by gamete 2.

One must also add the assumptions of independence of X_1 and X_2 and of Y_1 and Y_2. We are required to compute the genotypic distribution of the second generation, i.e.,

$$p = P(Y_1 = A, Y_2 = A)$$

$$q = P(Y_1 = a, Y_2 = a)$$

$$2r = P(Y_1 = A, Y_2 = a \text{ or } Y_1 = a, Y_2 = A).$$

We start with the computation of p. In view of the independence of Y_1 and Y_2, $p = P(Y_1 = A)P(Y_2 = A)$. By the rule of exclusive and exhaustive causes, $P(Y_1 = A) = P(Y_1 = A|X_1 = AA)P(X_1 = AA) + P(Y_1 = A|X_1 = A_a)P(X_1 = A_a) + P(Y_1 = A|X_1 = aa)P(X_1 = aa) = 1 \cdot x + \tfrac{1}{2} \cdot 22 + 0 \cdot y = x + z$. Therefore,

$$p = (x + z)^2,$$

and symmetry,

$$q = (y + z)^2.$$

Now $2r = P(Y_1 = A, Y_2 = a) + P(Y_1 = a, Y_2 = A)$, and therefore by symmetry, $r = P(Y_1 = A, Y_2 = a)$. In view of the independence of Y_1 and Y_2, $r = P(Y_1 = A)P(Y_2 = a)$. Finally, in view of previous computations,

$$2r = 2(x + z)(y + z).$$

Numerical Applications.

$$x = y = 2z = \frac{1}{3} \qquad \Rightarrow p = q = \frac{1}{4}, \quad 2r = \frac{1}{2}$$

$$x = \frac{1}{4}, \quad y = \frac{1}{2}, \quad 2z = \frac{1}{4} \Rightarrow p = \frac{9}{64}, \quad q = \frac{25}{64}, \quad 2r = \frac{30}{64}$$

$$x = y = \frac{1}{2}, \quad z = 0 \qquad \Rightarrow p = \frac{1}{4}, \quad q = \frac{1}{4}, \quad 2r = \frac{1}{2}.$$

E20. Define the functions $f_1, f_2,$ and f_3 by

$$f_1(x, y, z) = (x + z)^2$$

$$f_2(x, y, z) = (y + z)^2$$

$$f_3(x, y, z) = (x + z)(y + z).$$

To be proven: for all nonnegative numbers x, y, z such that $x + y + 2z = 1$,

$$f_i(x, y, z) = f_i[f_1(x, y, z), f_2(x, y, z), f_3(x, y, z)], \qquad i = 1, 2, 3.$$

The third equality, for instance, is

$$(x + z)(y + z) = [(x + z)^2 + (y + z)(x + z)][(y + z)^2 + (y + z)(x + z)].$$

It holds since $(x + z)^2 + (y + z)(x + z) = (x + z)(x + 2z + y) = x + z$ and $(y + z)^2 + (y + z)(x + z) = (y + z)(y + 2z + x) = y + z$. The ratio c of alleles of type A in the initial population is $x + z$. Now $y + z = 1 - c$. Therefore, the stationary distribution is

$$p = c^2, \qquad q = (1 - c)^2, \qquad 2r = 2c(1 - c).$$

CHAPTER 2

Discrete Probability

1. Discrete Random Elements

1.1. Discrete Probability Distributions

Let E be a denumerable set (i.e., finite or countable) and let $(\Omega, \mathscr{F}, P)$ be a probability space. Any function X mapping Ω into E and such that for all $x \in E$,

$$\{\omega \mid X(\omega) = x\} \in \mathscr{F},$$ (1)

is called a *discrete random element* of E. When $E \subset \mathbb{R}$, one would rather refer to X as a *discrete random variable*.

Requirement (1) allows us to define

$$p(x) = P(X = x).$$ (2)

The collection $[p(x), x \in E]$ is the *distribution of* X. It satisfies (see Eq. (18), Chapter 1)

$$0 \leqslant p(x) \leqslant 1, \sum_{x \in E} p(x) = 1.$$ (3)

EXAMPLE 1 (Single Toss of a Coin). The coin tossing experiment of a single coin with bias $p(0 \leqslant p \leqslant 1)$ is described by a discrete random variable X

taking its values in $E = \{0, 1\}$ with the distribution

$$P(X = 1) = p, \qquad P(X = 0) = q = 1 - p.$$

EXAMPLE 2 (A Finite Succession of Coin Tosses). Consider the probabilistic model $(\Omega, \mathscr{F}, P)$ of Chapter 1, Section 7. The sequence $(X_n, n \geq 1)$ is a sequence of independent random variables of the type described in Example 1 above. For each n, one can define a random element with values in $E = \{0, 1\}^n$, namely, $X = (X_1, \ldots, X_n)$.

E1 Exercise. What is the distribution of the random element X in Example 2?

EXAMPLE 3 (The Binomial Distribution). The setting is the same as in Example 2. The discrete random variable

$$S_n = X_1 + \cdots + X_n \tag{4}$$

takes its values in $E = \{0, 1, \ldots, n\}$. Its distribution was obtained in Chapter 1, Exercise E11, and it was found that $P(S_n = k) = p_k \ (0 \leq k \leq n)$ where

$$\boxed{p_k = \frac{n!}{k!(n-k)!} p^k q^{n-k}} \qquad (0 \leq k \leq n). \tag{5}$$

Distribution (5) is called the *binomial distribution* of size n and parameter p. Any discrete random variable Z with values in $E = \{0, \ldots, n\}$ and admitting this distribution is called a *binomial random variable* (of size n and parameter p). This is denoted by $Z \sim \mathscr{B}(n, p)$.

EXAMPLE 4 (The Geometric Distribution). Using the same definitions and notations as in Examples 2 and 3 above, define the random variable T to be the first time n for which $X_n = 1$ (Fig. 1). If no such n exists, $T = \infty$. Formally,

$$T = \begin{cases} \inf\{n \mid X_n = 1\} \text{ if } \{n \mid X_n = 1\} \neq \varnothing \\ +\infty \text{ otherwise.} \end{cases} \tag{6}$$

The random variable T therefore takes its values in $\bar{\mathbb{N}}_+$.

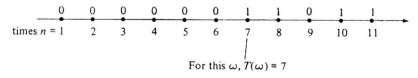

Figure 1. The geometric random variable.

E2 Exercise. Show that

$$\boxed{P(T = k) = pq^{k-1}} \qquad (k \geqslant 1). \tag{7}$$

Prove that $P(T = \infty) = 0$ or 1 according to whether $p > 0$ or $p = 0$.

One says that T admits a *geometric distribution* of parameter p, or equivalently, T is a *geometric random variable* of parameter p (see Fig. 2). This is symbolized by $T \sim \mathscr{G}(p)$.

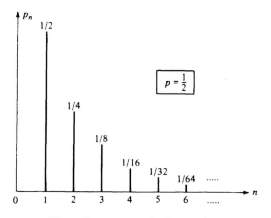

Figure 2. A geometric distribution.

E3 Exercise. Show that the geometric distribution has *no memory*, i.e., for all $n_0 \geqslant 1$,

$$\boxed{P(T \geqslant n_0 + k \mid T > n_0) = P(T \geqslant k)} \qquad (k \geqslant 1). \tag{8}$$

EXAMPLE 5 (The Multinomial Distribution). Suppose you have k boxes in which you place n balls at random in the following manner. The balls are thrown into the boxes independently of one another, and the probability that a given ball falls in box i is p_i (Fig. 3). Of course,

$$\boxed{0 \leqslant p_i \leqslant 1, \qquad \sum_{i=1}^{k} p_i = 1}. \tag{9}$$

Let X_i be the number of balls found in box i at the end of this procedure. The

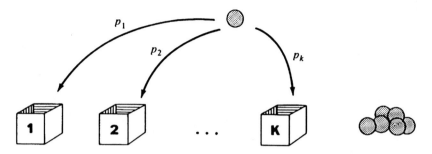

Figure 3. Placing balls at random in boxes.

random element $X = (X_1, \ldots, X_k)$ takes its values in the finite set E consisting of the k-tuples of integers $(m_1, \ldots, m_k)$ satisfying

$$m_1 + \cdots + m_k = n \qquad (10)$$

E4 Exercise. Show that $P(X_1 = m_1, \ldots, X_k = m_k) = p(m_1, \ldots, m_k)$ where

$$p(m_1, \ldots, m_k) = \frac{n!}{m_1! \ldots m_k!} p_1^{m_1} \ldots p_k^{m_k} \qquad (11)$$

Examine the case $k = 2$.

The probability distribution defined by Eqs. (9), (10), and (11) is the *multinomial distribution* of size (n, k) and of parameters $(p_1, \ldots, p_k)$. Notation $(X_1, \ldots, X_k) \sim \mathcal{M}(n, k, P_i)$ expresses that $(X_1, \ldots, X_k)$ is a multinomial random variable, i.e., admits a multinomial distribution.

EXAMPLE 6 (The Poisson Distribution). A random variable X that takes its value in $E = \mathbb{N}$ and admits the distribution

$$P(X = k) = e^{-\lambda} \frac{\lambda^k}{k!} \qquad (k \geq 0) \qquad (12)$$

where λ is a nonnegative real number, is called a *Poisson random variable* with parameter λ. This is denoted by $X \sim \mathcal{P}(\lambda)$. We recall the convention $0! = 1$ so that $P(X = 0) = e^{-\lambda}$ (see Fig. 4).

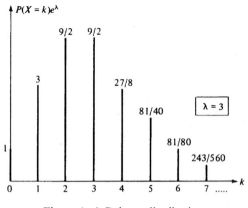

Figure 4. A Poisson distribution.

1.2. Expectation

The Real Case. Let X be a random element taking its values in E, with distribution $(p(x), x \in E)$, and let f be a function from E into $\mathbb{R}$. Suppose moreover that

$$\sum_{x \in E} |f(x)| p(x) < \infty. \tag{13}$$

One then defines the *expectation of* $f(X)$, denoted $E[f(X)]$, by

$$\boxed{E[f(X)] = \sum_{x \in E} f(x) p(x)}. \tag{14}$$

EXAMPLE 7. Let X be a Poisson random variable with parameter λ, and let f be the identity, i.e., $f(x) = x$. We have

$$E[X] = \sum_{k=0}^{\infty} k e^{-\lambda} \frac{\lambda^k}{k!}$$

$$E[X^2] = \sum_{k=0}^{\infty} k^2 e^{-\lambda} \frac{\lambda^k}{k!}.$$

E5 Exercise. Show that if $X \sim \mathscr{P}(\lambda)$, $E(X) = \lambda$ and $E[X^2] = \lambda + \lambda^2$.

EXAMPLE 8. Let $X = (X_1, \ldots, X_k)$ be the random element of Example 5, admitting the multinomial distribution described by Eqs. (9), (10), and (11). Take f to be the "projection" $f(x) = x_i$ where $1 \leq i \leq k$. Then

$$E[f(X)] = E[X_i] = \sum_{(m_1, \ldots, m_k) \in E} m_i \frac{n!}{m_1! \ldots m_k!} p_1^{m_1} \ldots p_k^{m_k}$$

where E is the set of k-tuples of integers $(m_1, \ldots, m_k)$ such that $\sum_{i=1}^k m_i = n$. We will not carry out the computation at this point (see Exercise E21).

Condition (13) is not the weakest one under which $E[f(X)]$ can be defined. When f is nonnegative, for instance, the series in the right-hand side of Eq. (14) has a meaning whether or not Eq. (13) is satisfied. However, if Eq. (13) is not required, $E[f(X)]$ can be infinite. When Eq. (13) is satisfied, $E[f(X)]$ is finite.

The Complex Case. Now let f be a function from E into $\mathbb{C}$ (the complex numbers) of the form

$$f(x) = g(x) + ih(x) \tag{15}$$

where $i = \sqrt{-1}$ and g and h are real valued functions such that $\sum_{x \in E} |g(x)| p(x) < \infty$ and $\sum_{x \in E} |h(x)| p(x) \, dx < \infty$, or, equivalently

$$\sum_{x \in E} |f(x)| p(x) \, dx < \infty.$$

The expectation of $f(X)$ is then defined by

$$\boxed{E[f(X)] = E[g(X)] + iE[h(X)]} \tag{16}$$

EXAMPLE 9. Let X be a Poisson random variable with parameter λ, and let $f(x) = s^x$ for some $s \in \mathbb{C}$. We have

$$E[f(X)] = E[s^X] = \sum_{k=0}^{\infty} s^k e^{-\lambda} \frac{\lambda^k}{k!}.$$

E6 Exercise. Carry out the computation in the Example 9 above to find $E[s^X] = e^{\lambda(s-1)}$. (This is the "generating" function of the Poisson distribution. More details in Section 3.)

Remark (A Question of Consistency). Consider a function f mapping E into $\mathbb{R}$, and define

$$Y = f(X). \tag{17}$$

Random variable Y is a discrete random variable taking its value in $F = f(E)$. Indeed $\{\omega \mid Y(\omega) = y\} = \{\omega \mid f(X(\omega)) = y\} = \bigcup_{x \in A_y} \{\omega \mid X(\omega) = x\}$ where A_y is the set of $x \in E$ such that $f(x) = y$, and therefore, since $\{\omega \mid X(\omega) = x\} \in \mathscr{F}$ for all $x \in E$, $\{\omega \mid Y(\omega) = y\} \in \mathscr{F}$. Suppose that condition (13) is satisfied. The question is, can we define $E[Y]$ by formula $E[Y] = \sum_{y \in F} yq(y)$ where $[q(y), y \in F]$ is the distribution of Y, and do we have

$$\sum_{x \in E} xp(x) = \sum_{y \in F} yq(y) \tag{18}$$

as expected?

E7 Exercise. Prove that the above question has a positive answer.

Properties of Expectation. One of the main properties, *linearity*, is obvious from the definitions. It says that if f_1 and f_2 are functions from E into $\mathbb{C}$ such that condition (13) is satisfied for $f = f_1$ and $f = f_2$, and if λ_1 and λ_2 are two complex numbers, then

$$E[\lambda_1 f_1(X) + \lambda_2 f_2(X)] = \lambda_1 E[f_1(X)] + \lambda_2 E[f_2(X)]. \tag{19}$$

The other important property of expectation is *monotonicity*. It says that if f_1 and f_2 above are restricted to take *real* values, and if

$$f_1(x) \leqslant f_2(x) \qquad (x \in E), \tag{20}$$

then

$$E[f_1(X)] \leqslant E[f_2(X)]. \tag{21}$$

The reader will easily provide the proof, and will see that Eq. (20) is required to hold only for those $x \in E$ such that $p(x) > 0$.

By the extension of the triangle inequality to series, we have

$$\left| \sum_{x \in E} f(x)p(x) \right| \leqslant \sum_{x \in E} |f(x)p(x)| = \sum_{x \in E} |f(x)|p(x),$$

that is,

$$|E[f(X)]| \leqslant E[|f(X)|], \tag{22}$$

an inequality frequently used.

Remark. The linearity property of expectation usually appears in the following form. For two discrete random variables X_1 and X_2,

$$E[X_1 + X_2] = E[X_1] + E[X_2]. \tag{23}$$

There is nothing new here; it suffices to consider the random discrete element $X = (X_1, X_2)$ and the functions f_1 and f_2 defined by $f_1(x) = x_1$ and $f_2(x) = x_2$ for all $x = (x_1, x_2)$. Then Eq. (23) is just

$$E[f_1(X) + f_2(X)] = E[f_1(X)] + E[f_2(X)].$$

There is a little trick often used in computations that is worthwhile mentioning at this point. Consider a random element X taking its values in E, and let C be a subset of E. The indicator function of C, denoted 1_C, is a function from E into $\mathbb{R}$ defined by

$$1_C(x) = \begin{cases} 1 & \text{if } x \in C \\ 0 & \text{if } x \notin C. \end{cases} \tag{24}$$

Applying definition (14) to $f = 1_C$, one obtains

$$E[1_C(X)] = \sum_{x \in E} 1_C(x)p(x) = \sum_{x \in C} p(x).$$

But

$$\sum_{x \in C} p(x) = \sum_{x \in C} P(X = x) = P\left(\sum_{x \in C} \{X = x\}\right) = P(X \in C).$$

Therefore,

$$\boxed{E[1_C(X)] = P(X \in C)}. \tag{25}$$

The above remark and the linearity and monotonicity properties of expectation will now be applied to derive a famous inequality.

Markov's Inequality. Let f be a function from E into $\mathbb{R}$ satisfying Eq. (13). Then,

$$\boxed{P(|f(X)| \geq a) \leq \frac{E[|f(X)|]}{a}} \qquad (a > 0). \tag{26}$$

PROOF OF EQ. (26). Let $C = \{x \,|\, |f(x)| \geq a\}$. Then $|f(x)| = 1_C(x)|f(x)| + 1_{\bar{C}}(x)|f(x)| \geq 1_C(x)|f(x)| \geq 1_C(x)a$ where the last inequality is obtained by noting that if $x \in C$ then $|f(x)| \geq a$, by definition of C. Therefore, using successively the monotonicity, the linearity, and the trick of Eq. (25),

$$E[|f(X)|] \geq E[a \cdot 1_C(X)] = aE[1_C(X)] = aP(X \in C) = aP(|f(X)| \geq a),$$

since $X \in C$ is equivalent to $|f(X)| \geq a$ by definition of C. $\qquad \square$

The indicator function trick (25) can be generalized as follows. Let $(\Omega, \mathscr{F}, P)$ be a probabilistic model and let A be some event. The indicator function 1_A is a function from Ω into $\{0, 1\}$ defined by

$$1_A(\omega) = \begin{cases} 1 & \text{if } \omega \in A \\ 0 & \text{if } \omega \notin A. \end{cases} \tag{27}$$

Clearly $X = 1_A$ is a discrete random variable taking its values in $E = \{0, 1\}$, and since the events $\{\omega \,|\, X(\omega) = 1\}$ and A are identical, $P(X = 1) = P(A)$. Similarly, $P(X = 0) = P(\bar{A}) = 1 - P(A)$. The expectation of X is

$$E[X] = 1 \cdot P(X = 1) + 0 \cdot P(X = 0) = P(X = 1) = P(A),$$

that is,

$$E[1_A] = P(A).$$ (28)

Identity (25) is a particular case of Eq. (28) with $A = \{X \in C\}$.

E8 Exercise (Inclusion–Exclusion Formula). Let $A_1, \ldots, A_n$ be n arbitrary events. Then

$$P\left(\bigcup_{k=1}^{n} A_k\right) = \sum_{i=1}^{n} P(A_i) - \sum_{\substack{i=1 \\ i<j}}^{n} \sum_{j=1}^{n} P(A_i \cap A_j)$$
$$+ \cdots + (-1)^{n+1} P\left(\bigcap_{k=1}^{n} A_k\right)$$ (29)

where the general term is

$$(-1)^{m+1} \sum_{\substack{i_1=1 \\ i_1<i_2<\cdots<i_m}}^{n} \cdots \sum_{i_m=1}^{n} P(A_{i_1} \cap \cdots \cap A_{i_m}).$$

Prove this formula using indicator functions. *Hint*: You will have to use the following identities:

$$\bigcap_{i=1}^{k} 1_{A_i} = \prod_{i=1}^{k} 1_{A_i}, \qquad 1_{\bar{A}} = 1 - 1_A.$$

1.3. Independence

The Product Formula. Let X and Y be two random elements with values in the denumerable spaces E and F respectively. Another random element Z taking its values in the denumerable space $G = E \times F$ can be constructed from X and Y by $Z = (X, Y)$.

Let $(p(x), x \in E)$, $(q(y), y \in F)$ and $(r(z), z \in G)$ be the distributions of X, Y, and Z, respectively. The random elements X and Y are said to be independent if and only if

$$r(z) = r(x, y) = p(x)q(y) \qquad (x \in E, y \in F, z = (x, y)),$$ (30)

that is,

$$P(X = x, Y = y) = P(X = x)P(Y = y) \qquad (x \in E, y \in F).$$ (31)

Now let f and g be two functions from E to $\mathbb{C}$ and from F to $\mathbb{C}$ respectively such that $\sum_{x \in E} |f(x)| p(x) < \infty$ and $\sum_{y \in F} |g(y)| q(y) < \infty$.

E9 Exercise. Show that $E[f(X)g(Y)]$ can be defined and that

$$E[f(X)g(Y)] = E[f(X)]E[g(Y)] \tag{32}$$

Equality (32) is called the *product formula*.

The Convolution Formula. Suppose that X and Y are discrete random variables taking their values in $\mathbb{N}$ and admitting the distributions

$$P(X = k) = p_k, \qquad P(Y = k) = q_k \qquad (k \geqslant 0).$$

If X and Y are independent, the random variable $S = X + Y$ admits the distribution

$$P(S = k) = r_k \qquad (k \geqslant 0)$$

defined by the *convolution formula*

$$r_k = \sum_{j=0}^{k} p_j q_{k-j} \tag{33}$$

PROOF OF EQ. (33). Since $\{S = k\} = \sum_{j=0}^{k}\{X = j, Y = k - j\}$, $r_k = \sum_{j=0}^{k} P(X = j, Y = k - j)$. But X and Y are independent and therefore

$$P(X = j, Y = k - j) = P(X = j)P(Y = k - j). \qquad \square$$

E10 Exercise. Let X and Y be two independent Poisson random variables with parameters λ and μ respectively. Show that $S = X + Y \sim \mathscr{P}(\lambda + \mu)$.

The independence concept can be generalized in a straightforward manner to an arbitrary number of discrete random elements $X_1, \ldots, X_n$ taking their values in $E_1, \ldots, E_n$, respectively. They are said to be independent if and only if

$$P(X_1 = x_1, \ldots, X_n = x_n) = P(X_1 = x_1) \ldots P(X_n = x_n)$$
$$(x_i \in E_i, 1 \leqslant i \leqslant n). \tag{34}$$

The product formula then takes the following form

$$E\left[\prod_{i=1}^{n} f_i(X_i)\right] = \prod_{i=1}^{n} E[f_i(X_i)], \tag{35}$$

where for each i, f_i is a mapping from E_i into $\mathbb{C}$ such that $\sum_{x_i \in E_i} |f_i(x_i)| P(X_i = x_i) < \infty$. The proof is the same as in Exercise E9.

2. Variance and Chebyshev's Inequality

2.1. Mean and Variance

If X is a discrete random variable, the quantities

$$m = E[X] \tag{36}$$

and

$$\sigma^2 = E[(X - m)^2] \tag{37}$$

are called (when they are defined) the *mean* and the *variance* of X, respectively, and σ ($\sigma \geq 0$) is the *standard deviation* of X.

The variance σ^2 can be obtained from the mean m and the *second moment* $E[X^2]$ as follows:

$$\sigma^2 = E[X^2] - m^2. \tag{38}$$

Indeed, by linearity, $E[(X - m)^2] = E[X^2] - 2mE[X] + m^2 = E[X^2] - 2m^2 + m^2$.

E11 Exercise. Using the results of Exercise E5, show that the variance of a Poisson random variable of parameter λ is λ.

E12 Exercise. Show that the mean and variance of a geometric random variable of parameter $p > 0$ are $1/p$ and q/p^2, respectively. (See Exercise E2 for the distribution of such a random variable.)

E13 Exercise. Let X be a discrete random variable with values in $\mathbb{N} = \{0, 1, 2, \ldots\}$. Show that

$$E[X] = \sum_{n=1}^{\infty} P(X \geq n). \tag{39}$$

E14 Exercise. Let $X_1, \ldots, X_m$ be m independent random variables with values in $\mathbb{N}$ and with a common distribution $P(X_i = k) = p_k (k \geq 0)$. Define $r_n = \sum_{k=n}^{\infty} p_k$. Using Eq. (39), show that

$$E[\min(X_1,\ldots,X_m)] = \sum_{n=1}^{\infty} r_n^m.$$

Apply this to the case $X_i \sim \mathscr{G}(p)$ (geometric).

Some Elementary Remarks. From the linearity of expectation it is clear that if X is a discrete random variable with mean m and variance σ^2, then the mean and variance of the random variable aX (where $a \in \mathbb{R}$) are am and $a^2\sigma^2$, respectively. Also if X has a vanishing variance ($\sigma^2 = 0$) then $P(X = m) = 1$. Indeed, $\sum_{x \in E}(x - m)^2 p(x) = 0$ implies that if $x \neq m$, $p(x) = 0$. Therefore, $P(X = m) = 1 - P(X \neq m) = 1 - \sum_{x \neq m} p(x)$. We shall see later that this property extends to arbitrary random variables (not necessarily discrete).

The Variance of a Sum of Independent Random Variables. Let $X_1, \ldots, X_n$ be n discrete random variables with variances $\sigma_1^2, \ldots, \sigma_n^2$, respectively. In the case where $X_1, \ldots, X_n$ are *independent*, the variance σ^2 of the sum

$$S = X_1 + \cdots + X_n$$

is equal to the sum of the variances:

$$\boxed{\sigma^2 = \sigma_1^2 + \cdots + \sigma_n^2}. \tag{40}$$

PROOF OF EQ. (40). Let m_i be the mean of X_i. Then $E[S] = \sum_{i=1}^{n} m_i$ and $\sigma^2 = E[S^2] - (\sum_{i=1}^{n} m_i)^2$. Also $E[S^2] = E[(\sum_{i=1}^{n} X_i)^2] = \sum_{i=1}^{n} E[X_i]^2 + 2\sum_{i=1}^{n}\sum_{j=1}^{n}_{i<j} E[X_i X_j] = \sum_{i=1}^{n} E[X_i^2] + 2\sum_{i=1}^{n}\sum_{j=1}^{n}_{i<j} m_i m_j$ where we have used the independence assumption. Now $2\sum_{i=1}^{n}\sum_{j=1}^{n} m_i m_j - (\sum_{i=1}^{n} m_i)^2 = -\sum_{i=1}^{n} m_i^2$. Therefore, $\sigma^2 = \sum_{i=1}^{n}(E[X_i^2] - m_i^2) = \sum_{i=1}^{n} \sigma_i^2$. □

E15 Exercise. Use Eq. (40) to show that the mean and variance corresponding to the binomial distribution of size n and parameter p are np and npq, respectively.

E16 Exercise. Let $X_1, \ldots, X_n$ be independent discrete random variables with common variance σ^2. Show that the *standard deviation* of the *empirical mean* $(X_1 + \cdots + X_n)/n$ is $\sigma/\sqrt{n}$.

2.2. Chebyshev's Inequality

When specialized to $f(x) = (x - m)^2$ and $a = \varepsilon^2$ where $\varepsilon > 0$, Markov's inequality (26) yields one of the most frequently used tool of Probability Theory, Chebyshev's inequality:

$$P(|X - m| \geq \varepsilon) \leq \frac{\sigma^2}{\varepsilon^2} \qquad (\varepsilon > 0) \qquad \qquad (41)$$

The Weak Law of Large Numbers. Let $(X_n, n \geq 1)$ be a sequence of discrete random variables, identically distributed with common mean m and common variance σ^2. Suppose, moreover, that they are independent (i.e., any finite collection $X_{i_1}, \ldots, X_{i_k}$ forms a collection of independent random variables). Consider the empirical mean

$$\frac{S_n}{n} = \frac{X_1 + \cdots + X_n}{n}. \qquad (42)$$

The mean of S_n/n is m, and as we saw in Exercise E16, its variance is σ^2/n. Application of Chebyshev's inequality to S_n/n yields

$$P\left(\left|\frac{S_n}{n} - m\right| \geq \varepsilon\right) \leq \frac{\sigma^2}{n\varepsilon^2}. \qquad (43)$$

Therefore, for all $\varepsilon > 0$,

$$\lim_{n\uparrow\infty} P\left(\left|\frac{S_n}{n} - m\right| \geq \varepsilon\right) = 0 \qquad (44)$$

This is the *weak law of large numbers.* It says that the empirical mean converges in probability to the probabilistic mean, according to the following definition of convergence in probability. A sequence of random variables $(X_n, n \geq 1)$ is said to *converge in probability* to a random variable X iff for all $\varepsilon > 0$

$$\lim_{n\uparrow\infty} P(|X_n - X| \geq \varepsilon) = 0 \qquad (45)$$

In Chapter 5, various notions of convergence will be introduced: convergence in quadratic mean, convergence in law, convergence in probability, and almost-sure convergence. In the hierarchy of convergence, almost-sure convergence implies convergence in probability. The *strong law of large numbers* states that the convergence of S_n/n to m takes place almost surely. The precise statement for this is as follows: for all $\omega \in \Omega$, apart from those in a set (event) N such that $P(N) = 0$, $\lim_{n\uparrow\infty}[S_n(\omega)/n] = m$. (The latter limit is in the ordinary sense.) The proof of the strong law of large numbers is not within our grasp at this point; it will be given in Chapter 5. However, it is interesting to note that Chebyshev's inequality will play a decisive role.

E17 Exercise. Using some physical apparatus, you measure a quantity, the actual value of which is unknown to you. Each measurement you make is equal to the actual value plus a random error (which can be negative). The errors in successive measurements are independent and have the same distribution. All you know about the error in a given experiment is that it has mean 0 and a standard deviation bounded by 10^{-4}. As usual, you perform n experiments, yielding n values (approximating up to errors the actual value), and you take the arithmetic mean to be the *experimental value* of the measured quantity. You want the difference between the experimental and the actual value to be less than 10^{-4} with probability larger than 0.99. What number n of experiments do you suggest?

E18 Exercise. Prove *Bernstein's polynomial approximation* of a continuous function f from $[0, 1]$ into $\mathbb{R}$, namely,

$$f(x) = \lim_{n \uparrow \infty} P_n(x) \qquad (x \in [0, 1]) \tag{46}$$

where

$$P_n(x) = \sum_{k=0}^{n} f\left(\frac{k}{n}\right) \frac{n!}{k!(n-k)!} x^k (1-x)^{n-k}, \tag{47}$$

and the convergence in Eq. (46) is *uniform* in $[0, 1]$. *Hint:* Consider a sequence $(X_n, n \geq 1)$ of random variables independent and identically distributed according to $P(X_n = 1) = x, P(X_n = 0) = 1 - x$, and compute $E[f(S_n/n)]$ where $S_n = X_1 + \cdots + X_n$.

3. Generating Functions

3.1. Definition and Basic Properties

The concept of generating function applies to discrete random variables with values in $E = \mathbb{N}$ and to discrete random elements with values in $E = \mathbb{N}^k$ for some $k > 1$. We begin with the univariate case $E = \mathbb{N}$.

Univariate Generating Function. Let X be a discrete random variable taking its values in $\mathbb{N}$ and admitting the distribution

$$P(X = k) = p_k \qquad (k \in \mathbb{N}).$$

The *generating function* of X is the function from the closed unit disc of $\mathbb{C}$ into $\mathbb{C}$ defined by

$$\boxed{g(s) = E[s^X]} \qquad (s \in \mathbb{C}, |s| \leq 1), \tag{48}$$

that is,

$$g(s) = \sum_{k=0}^{\infty} s^k P(X = k) = \sum_{k=0}^{\infty} s^k p_k. \tag{49}$$

Inside the unit disc, the power series $\sum p_k s^k$ is uniformly absolutely convergent since for $|s| < 1$

$$\sum_{k=1}^{\infty} p_k |s^k| \leqslant \sum_{k=1}^{\infty} p_k = 1.$$

One can therefore handle such series quite freely inside the unit disk ($|s| < 1$), for instance, add term by term and differentiate term by term. We will soon see how to make use of such possibilities.

The generating function characterizes the distribution in the following sense. If two random variables X_1 and X_2 taking their values in $E = \mathbb{N}$ have the same generating function:

$$g_1(s) = g_2(s) \qquad (|s| < 1), \tag{50}$$

then they have the same distribution:

$$P(X_1 = k) = P(X_2 = k) \qquad (k \in \mathbb{N}). \tag{51}$$

Indeed Eq. (50), is simply

$$\sum_{k=1}^{\infty} P(X_1 = k)s^k = \sum_{k=1}^{\infty} P(X_2 = k)s^k \qquad (|s| < 1),$$

and two power series that are convergent and identical in a neighborhood of 0 must have their corresponding coefficients equal.

E19 Exercise. Let X be a discrete random variable distributed according to the binomial distribution of size n and parameter $p[X \sim \mathscr{B}(n, p)]$. Show that the generating function of X is $g(s) = (ps + q)^n$.

E20 Exercise. Show that the generating function of a geometric random variable $X[X \sim \mathscr{G}(P)]$ is $g(s) = ps/(1 - qs)$.

Multivariate Generating Function. Let $X_1, \ldots, X_k$ be k discrete random variables taking their values in $\mathbb{N}$ and let

$$P(X_1 = i_1, \ldots, X_k = i_k) = p_{i_1 \ldots i_k}$$

be the distribution of the discrete random element $(X_1, \ldots, X_k)$. The generating function g of $(X_1, \ldots, X_k)$ is a mapping from $\mathbb{C}^k$ into $\mathbb{C}$ defined by

$$\boxed{g(s_1, \ldots, s_k) = E[s_1^{X_1} \ldots s_k^{X_k}]}, \tag{52}$$

that is,

$$g(s_1,\ldots,s_k) = \sum_{i_1=1}^{\infty} \cdots \sum_{i_k=1}^{\infty} s_1^{i_1}\ldots s_k^{i_k} P(X_1 = i_1,\ldots,X_k = i_k) \qquad (53)$$

or

$$g(s_1,\ldots,s_k) = \sum_{i_1=1}^{\infty} \cdots \sum_{i_k=1}^{\infty} s_1^{i_1}\ldots s_k^{i_k} p_{i_1\ldots i_k}.$$

Since when $|s_i| \leqslant 1$ for all $i(1 \leqslant i \leqslant k)$,

$$\sum_{i_1=1}^{\infty} \cdots \sum_{i_k=1}^{\infty} |s_1^{i_1}\ldots s_k^{i_k}| P(X_1 = i_1,\ldots,X_k = i_k)$$

$$\leqslant \sum_{i_1=1}^{\infty} \cdots \sum_{i_k=1}^{\infty} P(X_1 = i_1,\ldots,X_k = i_k) = 1$$

the domain of definition of g contains the subset of k-tuples $(s_1,\ldots,s_k)$ such that

$$|s_1| \leqslant 1,\ldots, |s_k| \leqslant 1. \qquad (54)$$

If $s_2 = \cdots = s_k = 1$,

$$E[s_1^{X_1} s_2^{X_2}\ldots s_k^{X_k}] = E[s_1^{X_1}] = g_1(s_1)$$

where g_1 is the generating function of X_1. Therefore,

$$g(s_1, 1,\ldots, 1) = g_1(s_1) . \qquad (55)$$

Similar relations hold for $X_2,\ldots, X_k$.

E21 Exercise. Find the generating function of $(X_1, X_2,\ldots, X_k) \sim \mathscr{M}(n, k, p_i)$. Show that $X_i \sim \mathscr{B}(n, p_i)$. Also, show that the variance of $X_1 + \cdots + X_n$ is not equal to the sum of the variances of the X_i's, as would be the case if the X_i's were independent.

Differentiation of Generating Functions and Moments. Since the power series $\sum_{k=0}^{\infty} p_k s^k$ converges absolutely when $|s| = 1$ ($\sum_{k=0}^{\infty} p_k |s|^k = \sum_{k=0}^{\infty} p_k = 1$), its radius of convergence is larger than or equal to 1, and inside the unit disk (i.e., for $|s| < 1$) one can differentiate term by term, at any order. Thus, for instance,

$$g'(s) = \sum_{k=1}^{\infty} k p_k s^{k-1} \qquad (|s| < 1).$$

When $s = 1$, the right-hand side of the above equality is $m = E[X] = \sum_{k=1}^{\infty} k p_k$. Therefore, by Abel's lemma

$$m = g'(1)$$ (56)

where $g'(1)$ is the limit of $g(s)$ when s tends to 1 with the constraint $|s| < 1$, $s \in \mathbb{R}$. Differentiating once more, one obtains $g''(s) = \sum_{k=1}^{\infty} k(k-1)p_k s^{k-2}$ and invoking Abel's lemma, $g''(1) = E[X^2] - m$, or, in the case where $m < \infty$, in view of Eq. (38),

$$\sigma^2 = g''(1) + g'(1) - g'(1)^2.$$ (57)

Here, $g''(1)$ admits an interpretation similar to $g'(1)$.

E22 Exercise. Compute the mean and variance of a binomial random variable $X [X \sim \mathscr{B}(n, p)]$ using generating functions and the result of Exercise E19.

E23 Exercise. Compute the mean and variance of a Poisson random variable $X [X \sim \mathscr{P}(\lambda)]$ using generating functions and the result of Exercise E6.

3.2. Independence and Product of Generating Functions

Let $X_1, \ldots, X_n$ be n discrete random variables with values in $\mathbb{N}$. Suppose that they are independent. Then by the product formula (35),

$$E[s_1^{X_1} \ldots s_n^{X_n}] = E[s_1^{X_1}] \ldots E[s_n^{X_n}],$$

that is,

$$g(s_1, \ldots, s_n) = g_1(s_1) \ldots g_n(s_n),$$ (58)

where g_i is the generating function of X_i and g is the generating function of $(X_1, \ldots, X_n)$. Now take $s_1 = \cdots = s_n = s$ in Eq. (58): $g(s, \ldots, s) = \prod_{j=1}^{n} g_i(s)$. But

$$g(s, \ldots, s) = E[s^{X_1} \ldots s^{X_n}] = E[s^{X_1 + \cdots + X_n}],$$

therefore $g(s, \ldots, s)$ is the generating function of $X_1 + \cdots + X_n$. We have thus obtained two results: if $X_1, \ldots, X_n$ are independent, Eq. (58) is true, and the generating function of the sum $X_1 + \cdots + X_n$ is equal to the product of the generating functions of $X_1, \ldots, X_n$.

E24 Exercise. Using generating functions, show that if X_1 and X_2 are independent Poisson random variables, $X_1 \sim \mathscr{P}(\lambda_1)$ and $X_2 \sim \mathscr{P}(\lambda_2)$, then $X_1 + X_2 \sim \mathscr{P}(\lambda_1 + \lambda_2)$.

Wald's Equality. Let $(X_n, n \geq 1)$ be a sequence of independent discrete random variables with values in $\mathbb{N}$ and identically distributed with the common generating function g_X. Now let T be a discrete random variable, with values in $\mathbb{N}$ and generating function g_T. Suppose, moreover, that T is independent of the X_n's. We are going to compute the generating function g_Y of

$$Y = X_1 + \cdots + X_T, \tag{59}$$

using the various tricks that we have gathered along the way.

First we write the definition

$$g_Y(s) = E[s^Y] = E[s^{X_1 + \cdots + X_T}].$$

From the identity $\Omega = \sum_{n=1}^{\infty} \{T = n\}$, (or, in terms of indicator functions, $1 = \sum_{n=1}^{\infty} 1_{\{T=n\}}$),

$$E[s^{X_1 + \cdots + X_T}] = E\left[\sum_{n=1}^{\infty} (1_{\{T=n\}} s^{X_1 + \cdots + X_T}) \right]$$

$$= E\left[\sum_{n=1}^{\infty} 1_{\{T=n\}} s^{X_1 + \cdots + X_n} \right],$$

where we have observed that if $T = n$, $X_1 + \cdots + X_T = X_1 + \cdots + X_n$. Now

$$E\left[\sum_{n=0}^{\infty} 1_{\{T=n\}} s^{X_1 + \cdots + X_n} \right] = \sum_{n=0}^{\infty} E[1_{\{T=n\}} s^{X_1 + \cdots + X_n}].$$

(We must confess that we have been abusing the linearity property, having used it for infinite sums. Later, in Chapter 3, you will find an excuse for this: it is called Lebesgue's dominated convergence theorem.) In view of the independence assumptions,

$$E[1_{\{T=n\}} s^{X_1 + \cdots + X_n}] = E[1_{\{T=n\}}] E[s^{X_1 + \cdots + X_n}]$$

$$= E[1_{\{T=n\}}] E[s^{X_1}] \ldots E[s^{X_n}] = P(T = n) g_X(s)^n.$$

Finally

$$E[s^Y] = \sum_{n=0}^{\infty} P(T = n) g_X(s)^n,$$

that is,

$$\boxed{g_Y(s) = g_T[g_X(s)]}. \tag{60a}$$

E25 Exercise. Under the above assumptions on $(X_n, n \geq 1)$ and T, prove Wald's equality:

$$\boxed{E\left[\sum_{n=1}^{T} X_n \right] = E[X_1] \cdot E[T]}. \tag{60b}$$

Illustration 4.   An Introduction to Population Theory: Galton–Watson's Branching Process

The English statistician Galton is considered to be the initiator of the theory of branching processes. He studied the transmission of family names through generations and was particularly interested in estimating the survival probability of a given branch in a genealogical tree. A particularly simple model of the situation he investigated is the following one.

All the individuals of a given colony (e.g., the male population of a noble family) give birth in their lifetime to a random number of descendents (e.g., the male descendents in the particular example of interest to Galton). Each individual of the colony procreates independently of all other members of the colony. If X_n denotes the size of the nth generation,

$$X_{n+1} = \begin{cases} \sum_{i=1}^{X_n} Z_n^{(i)} & \text{if } X_n \geq 1 \\ 0 & \text{if } X_n = 0, \end{cases} \tag{61}$$

where $(Z_n^{(i)}, i \geq 0, n \geq 0)$ are identically distributed independent random variables, integer valued, with common generating function

$$g_Z(s) = \sum_{n=0}^{\infty} P(Z = n)s^n, \tag{62}$$

with finite mean m and finite variance σ^2. In this model, the number X_0 of ancestors is naturally supposed to be independent of the random variables $(Z_n^{(i)}, i \geq 0, n \geq 0)$. This simple model is a particular case of *branching process* (Fig. 5 motivates the terminology) and was studied by the probabilist Watson, who gave the following analysis.

E26 Exercise. Denote by ϕ_n the generating function of X_n:

$$\phi_n(s) = \sum_{k=0}^{\infty} P(X_n = k)s^k. \tag{63}$$

Show that

$$\boxed{\phi_{n+1}(s) = \phi_n(g_Z(s))}. \tag{64}$$

Deduce from the above relation that if $X_0 = 1$ (one ancestor),

Illustration 4. An Introduction to Population Theory — 65

$$\phi_{n+1}(s) = g_Z(\phi_n(s)). \tag{65}$$

Compute the mean and variance of X_n in the case where $X_0 = 1$, and then in the case where $X_0 = k > 1$.

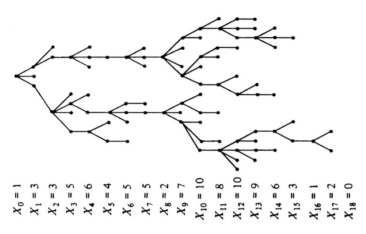

$X_0 = 1$
$X_1 = 3$
$X_2 = 3$
$X_3 = 5$
$X_4 = 6$
$X_5 = 4$
$X_6 = 5$
$X_7 = 5$
$X_8 = 2$
$X_9 = 7$
$X_{10} = 10$
$X_{11} = 8$
$X_{12} = 10$
$X_{13} = 9$
$X_{14} = 6$
$X_{15} = 3$
$X_{16} = 1$
$X_{17} = 2$
$X_{18} = 0$

Figure 5. A typical realization ω giving rise to a genealogical tree when there is one ancestor $(X_0(\omega) = 1)$. In this example, there is extinction of the family name at the 18th generation.

An extinction occurs if for some $n \geqslant 1$, $X_n = 0$, since in this case $X_{n+j} = 0$ for all $j \geqslant 0$. Therefore, denoting P_e as the probability of extinction,

$$P_e = P\left(\bigcup_{n=1}^{\infty} \{X_n = 0\}\right).$$

But $(\{X_n = 0\}, n \geqslant 1)$ is an increasing sequence of events since $X_n = 0$ implies $X_{n+1} = 0$, so that, by the sequential continuity property [Eq. (8) of Chapter 1],

$$P_e = \lim_{n \uparrow \infty} \uparrow P(X_n = 0).$$

Now, from Eq. (63), $P(X_n = 0) = \phi_n(0)$, and therefore by Eq. (65), $P(X_{n+1} = 0) = g_Z[P(X_n = 0)]$. Letting $n \uparrow \infty$ in the latter equality yields

$$\boxed{P_e = g_Z(P_e)}. \tag{66}$$

To find P_e we must therefore study the equation $x = g_Z(x)$, $x \in [0, 1]$. Clearly $g_Z'(x) = \sum_{n=1}^{\infty} nP(Z = n)x^{n-1} \geqslant 0$ for $x \in [0, 1]$ and therefore g_Z is non-decreasing in $[0, 1]$. If we exclude the trivial case where $P(Z = 0) = 1$, g_Z' is strictly positive in $(0, 1]$. Also $g_Z'(0) = P(Z = 1)$ and $g_Z'(1) = E[Z] = m$.

Differentiating once more, we see that $g_Z''(x) = \sum_{n=2}^{\infty} n(n-1)P(Z = n) \times x^{n-2} \geq 0$ in $[0, 1]$, so that g_Z is a $\cup$-convex function in $[0, 1]$.

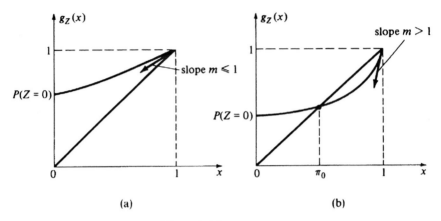

(a)   (b)

Figure 6. The two cases.

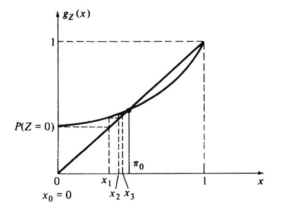

Figure 7. The iteration leads to π_0.

In summary, when $P(Z = 0) \in (0, 1)$, two cases arise. If $E[Z] \leq 1$ (Fig. 6a), $P_e = 1$ since this is the unique solution to Eq. (66) with $P_e \in [0, 1]$. If $E[Z] > 1$ (Fig. 6b), Eq. (66) has two solutions in $[0, 1]$, 1 and $\pi_0 < 1$. We will see that 1 must be excluded. Indeed, letting $x_n = P(X_n = 0)$, we have shown previously that $x_{n+1} = g_Z(x_n)(n \geq 0)$ where $x_0 = P(X_0 = 0) = 0$. It is clear from **Fig. 7** that the iteration scheme leads to π_0, not 1. The formal proof of this is an exercise in calculus, which is left to the reader.

E27 Exercise. Denote by $P_{e,k}$ the probability of extinction when $X_0 = k$ (k ancestors). What is the relation between $P_{e,k}$ and $P_{e,1} = P_e$?

Illustration 5. Shannon's Source Coding Theorem 67

Before closing this Illustration, let us mention that the theory of branching processes is a very fertile area of Applied Probability and that it finds applications not only in Sociology, but also in Epidemiology, Physics, and many other fields. For instance, consider a nuclear reactor in which heavy particles (neutrons) collide with molecules, extracting more particles at each step or being absorbed. This is a typical branching process. It is interesting to be able to find conditions that prevent extinction of the pile, and branching process theory has provided some answers to this problem.

Illustration 5.   Shannon's Source Coding Theorem: An Introduction to Information Theory

The American telecommunications engineer Claude Shannon discovered in 1948 an entirely new area of Applied Mathematics: Information Theory. The following is a guided introduction to this theory. We shall prove and use Shannon's *source coding theorem* which, together with Huffman's coding theorem, enables one to compress data (e.g., in digital form) using previous knowledge of the statistical distribution of such data.

Quantity of Information. Let $X_1, \ldots, X_n$ be discrete random elements taking their values in the finite set E, called an *alphabet*, with the generic element, or *letter*, α_i:

$$E = \{\alpha_1, \ldots, \alpha_k\}.$$

Denote $X = (X_1, \ldots, X_n)$. Therefore, X is a random element taking its values in the finite set E^n, the set of *words of length* n written in the alphabet E. For any random element Y taking its values in a finite set F, with the distribution

$$P(Y = y) = p(y),$$

one defines the average *quantity of information* $H(Y)$ contained in Y by

$$\boxed{H(Y) = -E[\log p(Y)]}. \tag{67}$$

The base of the logarithm will be specified only when needed. By convention, $0 \log 0 = 0$.

E28 Exercise. Compute $H(X_1)$ when $P(X_1 = \alpha_i) = p_i$. Also, compute $H(X) = H(X_1, \ldots, X_n)$ when $X_1, \ldots, X_n$ are iid.

E29 Exercise (Gibbs' Inequality). Using the well-known inequality

$$\log z \leqslant z - 1 \qquad (z > 0)$$

with equality if and only if $z = 1$, prove Gibbs' inequality

$$-\sum_{i=1}^{k} p_i \log p_i \leqslant -\sum_{i=1}^{k} p_i \log q_i \tag{68}$$

where $(p_i, 1 \leqslant i \leqslant k)$ and $(q_i, 1 \leqslant i \leqslant k)$ are arbitrary discrete probability distributions. Show that equality takes place if and only if

$$p_i = q_i \qquad (1 \leqslant i \leqslant k).$$

Deduce from Gibbs' inequality the inequality

$$H(X_1) \leqslant \log k \tag{69}$$

where equality holds if and only if

$$P(X_1 = \alpha_i) = p_i \equiv \frac{1}{k} \qquad (1 \leqslant i \leqslant k).$$

Show that

$$H(X_1) \geqslant 0 \tag{70}$$

with equality if and only if for some $j \in \{1, 2, \ldots, k\}$, $P(X_1 = \alpha_j) = 1$.

Coding and Kraft's Inequality for Prefix Codes. A binary *code for E* is a mapping c from E into $\{0, 1\}^*$ the set of finite sequences (including the empty one) of 0's and 1's. In this context $c(\alpha_i)$ is the code word for α_i. Let $l_i(c)$ be the length of this code word. Code c is said to be a *prefix code* if there exists no pair (i, j) with $i \neq j$ such that $c(\alpha_i)$ is the beginning of $c(\alpha_j)$.

EXAMPLES. $E = \{\alpha_1, \alpha_2, \alpha_3, \alpha_4\}$. The code

$$\alpha_1 \rightarrow c(\alpha_1) = 0$$

$$\alpha_2 \rightarrow c(\alpha_2) = 1$$

$$\alpha_3 \rightarrow c(\alpha_3) = 10$$

$$\alpha_4 \rightarrow c(\alpha_4) = 11$$

is not a prefix code since $c(\alpha_2)$ is the beginning of both $c(\alpha_3)$ and $c(\alpha_4)$. But

Illustration 5. Shannon's Source Coding Theorem 69

$$\alpha_1 \to c(\alpha_1) = 00$$
$$\alpha_2 \to c(\alpha_2) = 01$$
$$\alpha_3 \to c(\alpha_3) = 10$$
$$\alpha_4 \to c(\alpha_4) = 11$$

is a prefix code.

Prefix codes are *uniquely decodable*, i.e., if a finite sequence of 0's and 1's is obtained by encoding a finite string of letters from E, the original string of letters can be unambiguously reconstructed. For instance, the encoding of $\alpha_2\alpha_1$ using the first code yields 10, but this result is also obtained by encoding α_3 using the same code. Thus, the first code is not uniquely decodable.

Represent any binary code on the binary tree. For instance, the first code can be represented as in Fig. 8(a) and the second as in Fig. 8(b).

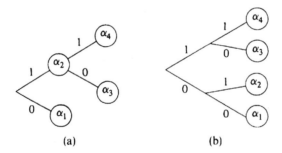

(a)   (b)

Figure 8. Two codes on a tree.

E30 Exercise. From the representations shown in Figs. 8(a) and 8(b) deduce that if c is a prefix code

$$\sum_{i=1}^{k} 2^{-l_i(c)} \leq 1.$$

Conversely, if a set of integers $(l_i, 1 \leq i \leq k)$ satisfies Kraft's inequality

$$\boxed{\sum_{i=1}^{k} 2^{-l_i} \leq 1}, \tag{71}$$

then there exists at least one binary prefix code c for E with $l_i(c) = l_i (1 \leq i \leq k)$. (*Note*: You can solve Exercise E31 using the answer to this question if you want to progress and go back to Exercise E30 later.)

E31 Exercise. Denote by $\mathscr{P}$ the set of binary prefix-codes for E. Define for any binary code c for E its average length $L(c)$ by

$$L(c) = \sum_{i=1}^{k} p_i l_i(c) \,. \tag{72}$$

Show that

$$H_2(X_1) \leqslant \inf_{c \in \mathscr{P}} L(c) \leqslant H_2(X_1) + 1 \tag{73}$$

where $H_2(X_1)$ is a notation for $H(X_1)$ when the base of the log used in definition (67) is 2. (*Hint*: Solve a minimization problem under Kraft's constraint and remember that the lengths $l_i(c)$ must be integers.)

Huffman's Code.

E32 Exercise. Suppose that $p_1 \geqslant \cdots \geqslant p_k$. Let c be a binary prefix code for E with $l_i(c) = l_i$. Consider the two following properties

$$(\mathscr{P}_1) \qquad l_1 \leqslant \cdots \leqslant l_k$$

$$(\mathscr{P}_2) \qquad l_k = l_{k-1}$$

Show that if they are not both satisfied by c, there exists a binary prefix code not worse than c that satisfies both of them (not worse = with smaller or equal average length).

Assuming $(\mathscr{P}_1)$ and $(\mathscr{P}_2)$, show that there exists a code not worse than c with the configuration shown in the figure below for the code words corresponding to $k - 1$ and k.

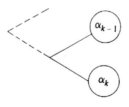

Deduce from the above remarks an algorithm for constructing a binary prefix code for E with minimal average length (*Hint*: Consider the set

$$E' = E - \{\alpha_{k-1}, \alpha_k\} + \{\alpha'_{k-1}\} = \{\alpha_1, \ldots, \alpha_{k-2}, \alpha'_{k-1}\}$$

with the probability distribution

Illustration 5. Shannon's Source Coding Theorem 71

$$p_1, p_2, \ldots, p_{k-2}, p'_{k-1} = p_{k-1} + p_k$$

and apply the above remarks to an optimal code for E'.) (*Note*: Exercise E34 requires the knowledge of Huffman's procedure.)

Block Encoding. Instead of encoding E we will encode E^n, the set of *blocks* of length n. Let the letters of the "words" (of length n) of E^n be chosen independently of one another and drawn from the alphabet E according to

$$P(X_j = \alpha_i) = p_i \qquad (1 \leqslant j \leqslant n, 1 \leqslant i \leqslant k).$$

E33 Exercise. Let $c^{(n)}$ be an optimal binary prefix code for E^n. Show that

$$\boxed{\lim_{n \uparrow \infty} \frac{L(c^{(n)})}{n} = H_2(X_1)}. \tag{74}$$

In other words, $H_2(X_1)$ is asymptotically the smallest average number of binary digits needed *per symbol* of E.

E34 Exercise. Consider a source that emits an infinite sequence $(X_n, n \geqslant 1)$ of $\{0, 1\}$-valued iid random variables with $P(X_n = 1) = p = \frac{3}{4}$ at the rate of one binary symbol every unit of time. Compute $H_2(p, 1 - p)$. Suppose that this sequence is transmitted via a channel that accepts only 0.83 binary symbol every unit of time. Give an explicit means of adaptation of the source to the channel.

The Questionnaire Interpretation. Suppose that one among k "objects" $\alpha_1, \ldots,$ α_k is drawn at random according to the probability distribution $p_1, \ldots, p_k$. This object is not shown to you and you are required to identify it by asking questions that can be answered by yes or no. Of course the k objects are distinguishable. What is the best you can do if you are allowed to ask any yes or no questions? What is the least average number of such questions needed to identify the object?

Clearly, since a yes or no question is associated with a partition into two classes, the first question will be associated with a partition of $E = \{\alpha_1, \ldots, \alpha_k\}$ into two classes E_1 and E_0 ($E_0 \cap E_1 = \varnothing$ and $E_0 + E_1 = E$). Suppose the answer is yes, i.e., "the object is in E_1." The next question will concern only E_1 for otherwise you lose time and your questionnaire is not optimal. This question will be associated with a partition $E_1 = E_{10} + E_{11}$. The same would have been true if the answer to the first question had been no, i.e., "the object is in E_0." The second question would then have been associated with a partition $E_0 = E_{00} + E_{01}$.

Of course $E_0, E_1, E_{00}, E_{01}, \ldots$ must be nonempty otherwise your questionnaire is not optimal. From the above argument, we see that a questionnaire which is a candidate to optimality results in a tree (Fig. 9). This tree is not

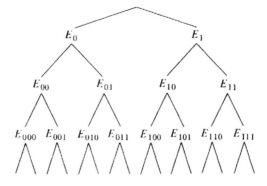

Figure 9. A questionnaire tree.

infinite: we stop the dichotomic procedure when a set at a node contains just one object because we then have identified the said object.

As you now see, a good questionnaire is associated with a prefix code for $E = \{x_1, \ldots, x_k\}$. Therefore, to find an optimal questionnaire, you have to construct the Huffman prefix code for E associated with the distribution $(p_1, \ldots, p_k)$.

EXAMPLE (The Coin Problem). Suppose you have 15 coins identical in all respects, except that one and only one among them has a weight slightly different from the others. The only means you have to identify the odd coin is a scale. Find an optimal weighing strategy.

We propose the following analysis. The coins are laid down in front of you (Fig. 10). You will call this configuration #11 because the odd coin is the 11th from the left. Now you must identify a configuration among the 15 possible equiprobable configurations. The scale is a yes or no answering device because it has three positions: balance, left, and right, but from the point of view of our search, left and right are the same because we do not know whether the odd coin is lighter or heavier than the others.

Figure 10. Coins.

We will therefore find the optimal yes or no questionnaire and then check that its partition procedure is implementable by the scale. If yes, we will certainly have found the best weighing strategy.

The binary Huffman code for 15 equiprobable objects is summarized in Fig. 11. The first yes or no question must discriminate between configurations 1, 2, 3, 4, 5, 6, 7, 8 and configurations 9, 10, 11, 12, 13, 14, 15. How can we achieve this with the scale? One possible answer is to weigh the four leftmost coins against the next four coins. If there is balance, the odd coin must be in position

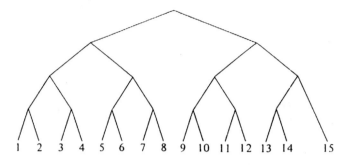

Figure 11. Optimal search for the slightly different coin among 15 coins.

9, 10, 11, 12, 13, 14, or 15. If not, it is in position 1, 2, 3, 4, 5, 6, 7, or 8. Note, however, that if the scale is unbalanced, one cannot determine whether the odd coin is in position 1, 2, 3, or 4 or if it is in position 5, 6, 7, or 8 because one does not know whether it is heavier or lighter than the ordinary coins.

The other steps are analogous to the first step except for the last one in case we have to make a final decision about two coins. In this situation, we just weigh one of the two coins against one of the coins that we have eliminated in the previous steps because such coins are normal.

In summary, the Huffman search strategy is in this case implementable.

E35 Exercise. There are six bags containing coins. All of them, except one, are filled with "good" coins. One of them, unknown to you, contains $\frac{1}{3}$ "bad" coins (slightly heavier or slightly lighter). You draw one coin from each bag and you want to determine if there is a bad coin among the six and if there is one, which one. Find an optimal search strategy.

SOLUTIONS FOR CHAPTER 2.

E1. For $x = (x_1, x_2, \ldots, x_n) \in \{0, 1\}^n$, $P(X = x) = P(X_1 = x_1, \ldots, X_n = x_n) = \prod_{i=1}^{n} P(X_i = x_i)$ (independence). Since $P(X_i = x_i) = p$ if $x_i = 1$, $P(X_i = x_i) = q$ if $x_i = 0$,

$$P(X = x) = p^{\sum_{i=1}^{n} x_i} q^{n - \sum_{i=1}^{n} x_i} \qquad (x \in \{0, 1\}^n).$$

E2. $P(T = k) = P(X_1 = 0, \ldots, X_{k-1} = 0, X_k = 1) = P(X_1 = 0) \ldots P(X_{k-1} = 0) \times P(X_k = 1) = q^{k-1} p$. Since $P(\{T = 1\} + \{T = 2\} + \cdots + \{T = \infty\}) = 1$, we have $P(T = \infty) = 1 - P(\sum_{k=1}^{\infty} \{T = k\}) = 1 - \sum_{k=1}^{\infty} P(T = k) = 1 - \sum_{k=1}^{\infty} pq^{k-1}$. But $q < 1$ when $p > 0$, and therefore $\sum_{k=1}^{\infty} pq^{k-1} = p/(1 - q) = p/p = 1$. If $p = 0$, $P(T = k) = 0$, and therefore $P(T = \infty) = 1 - \sum_{k=1}^{\infty} P(T = k) = 1$.

E3. The case $p = 0$ is easy and without interest. We therefore suppose $p > 0$. For all $m \geq 1$, $P(T \geq m) = p - q^{m-1}(1 + q + q^2 + \cdots) = pq^{m-1}/(1 - q) = q^{m-1}$. Therefore for $k \geq 1$, $P(T \geq n_0 + k | T > n_0) = P(T \geq n_0 + k, T > n_0)/P(T > n_0) = P(T \geq n_0 + k)/P(T \geq n_0 + 1) = q^{n_0 + k - 1}/q^{n_0} = q^{k-1} = P(T \geq k)$.

E4. If the balls are distinguishable (for instance, they are numbered from 1 to n), there
     are $n!/(m_1!\ldots m_k!)$ different possibilities of obtaining the configuration $m_1,\ldots,m_k$
     (i.e., m_1 balls in box 1, ..., m_k balls in box k), for instance, the m_1 first balls in 1,
     the m_2 following in 2, etc. Each possibility has probability $p_1^{m_1}\ldots p_k^{m_k}$ since the
     balls are placed independently of one another. Therefore, by the additivity axiom,
     the probability of configuration $(m_1,\ldots,m_k)$ is $[n!/(m_1!\ldots m_k!)]p_1^{m_1}\ldots p_2^{m_k}$.

     In the case where $k = 2$, we have $X_2 = n - X_1$, $p_2 = 1 - p_1 = q_1$. Therefore
     $P(X_1 = m_1) = P(X_1 = m_1, X_2 = n - m_1) = [n!/m_1!(n - m_1)!]p_1^{m_1}q_1^{n-m_1}$, where
     $0 \leqslant m_1 \leqslant n$. This is the binomial distribution.

E5.
$$E[X] = \sum_{k=1}^{\infty} ke^{-\lambda}\frac{\lambda^k}{k!} = \lambda e^{-\lambda}\sum_{k=1}^{\infty}\frac{\lambda^{k-1}}{(k-1)!} = \lambda e^{-\lambda}\sum_{k=1}^{\infty}\frac{d}{d\lambda}\left(\frac{\lambda^k}{k!}\right)$$

$$= \lambda e^{-\lambda}\frac{d}{d\lambda}\left(\sum_{k=0}^{\infty}\frac{\lambda^k}{k!}\right) = \lambda e^{-\lambda}\frac{d}{d\lambda}(e^{\lambda}) = \lambda e^{-\lambda}e^{\lambda} = \lambda.$$

$$E[X^2] = \sum_{k=0}^{\infty} k^2 \cdot e^{-\lambda}\frac{\lambda^k}{k!} = \sum_{k=1}^{\infty}k(k-1)e^{-\lambda}\frac{\lambda^k}{k!} + \sum_{k=1}^{\infty}ke^{-\lambda}\frac{\lambda^k}{k!}.$$

$$\sum_{k=1}^{\infty}k(k-1)e^{-\lambda}\frac{\lambda^k}{k!} = \lambda^2 e^{-\lambda}\sum_{k=1}^{\infty}\frac{d^2}{d\lambda^2}\left(\frac{\lambda^k}{k!}\right) = \lambda^2 e^{-\lambda}\frac{d^2}{d\lambda^2}\left(\sum_{k=0}^{\infty}\frac{\lambda^k}{k!}\right)$$

$$= \lambda^2 e^{-\lambda}\frac{d^2}{d\lambda^2}(e^{\lambda}) = \lambda^2 e^{-\lambda}e^{\lambda} = \lambda^2.$$

E6.
$$\sum_{k=0}^{\infty} s^k e^{-\lambda}\frac{\lambda^k}{k!} = e^{-\lambda}\sum_{k=0}^{\infty}\frac{(\lambda s)^k}{k!} = e^{-\lambda}e^{\lambda s} = e^{\lambda(s-1)}.$$

E7. The distribution of Y is

$$P(Y = y) = q(y) = \sum_{x \in A_y} p(x),$$

since $\{Y = y\} = \sum_{x \in A_y}\{X = x\}$. Therefore,

$$\sum_{y \in F}|y|q(y) = \sum_{y \in F}\left(|y|\sum_{x \in A_y}p(x)\right) = \sum_{y \in F}\left(\sum_{x \in A_y}|y|p(x)\right).$$

Since $x \in A_y$ is equivalent to $y = f(x)$, for fixed y, $\sum_{x \in A_y}|y|p(x) = \sum_{x \in A_y}|f(x)|p(x)$.
Therefore, using indicator functions $[1_A(x) = 1$ if $x \in A$, 0 otherwise],
$\sum_{y \in F}|y|q(y) = \sum_{y \in F}\sum_{x \in A_y}|f(x)|p(x) = \sum_{y \in F}\sum_{x \in E}1_{A_y}(x)|f(x)|p(x) = $
$\sum_{x \in E}\sum_{y \in F}1_{A_y}(x)|f(x)|p(x) = \sum_{x \in E}(\sum_{y \in F}1_{A_y}(x))|f(x)|p(x)$. But $\sum_{y \in F}1_{A_y}(x) = $
$1_{\bigcup_y A_y}(x) = 1_E(x) = 1$. Therefore,

$$\sum|y|q(y) = \sum|f(x)|p(x) < \infty.$$

One then proves in the same way that

$$\sum_{y \in F} yq(y) = \sum_{x \in E} f(x)p(x).$$

E8. First observe that $1_{B \cap C} = 1_B \cdot 1_C$ and that $1_{\bar{A}} = 1 - 1_A$. Therefore, the indicator
     function of $\bigcup_{k=1}^{n} A_k$ is 1 minus the indicator function of $\overline{\bigcup_{k=1}^{n} A_k}$, that is, by de
     Morgan's rule, 1 minus the indicator function of $\bigcap_{k=1}^{n} \bar{A}_k$, or $1 - \prod_{k=1}^{n}1_{\bar{A}_k} = $
     $1 - \prod_{k=1}^{n}(1 - 1_{A_k})$. Therefore, by Eq. (28) $P(\bigcup_{k=1}^{n} A_k) = E[1 - \prod_{k=1}^{n}(1 - 1_{A_k})]$.
     Letting $X_k = 1_{A_k}$, we have

$$P\left(\bigcup_{k=1}^{n} A_k\right) = E\left[1 - \prod_{k=1}^{n}(1 - X_k)\right]$$

$$= E\left[\sum_{i=1}^{n} X_i - \sum_{\substack{i=1 \\ i<j}}^{n}\sum_{j=1}^{n} X_i X_j + \cdots + \right.$$

$$\left. (-1)^m \sum_{\substack{i_1 \\ i_1<\cdots<i_m}}^{n}\cdots\sum_{i_m}^{n} X_{i_1}\ldots X_{i_m} + \cdots + (-1)^n X_1 \ldots X_n\right]$$

$$= \sum_{i=1}^{n} E[X_i] - \sum_{\substack{i=1 \\ i<j}}^{n}\sum_{j=1}^{n} E[X_i X_j] + \cdots + $$

$$(-1)^m \sum_{\substack{i_1=1 \\ i_1<\cdots<i_m}}^{n}\cdots\sum_{i_m=1}^{n} E[X_{i_1}\ldots X_{i_m}] + \cdots + (-1)^n E[X_1 \ldots X_n].$$

But

$$E[X_{i_1}\ldots X_{i_m}] = E[1_{A_{i_1}}\ldots 1_{A_{i_m}}] = E[1_{A_{i_1}\cap\cdots\cap A_{i_m}}] = P(A_{i_1} \cap \cdots \cap A_{i_m}).$$

E9. In order to show that $E[f(X)g(Y)]$ is defined, one must prove that $\sum_{z \in F}|h(z)|r(z) < \infty$, where $h(z) = f(x)g(y)$ and $z = (z, y)$. But

$$\sum_{z} |h(z)|r(z) = \sum_{x}\sum_{y} |f(x)||g(y)|p(x)q(y)$$

$$= \left(\sum_{x} |f(x)|p(x)\right)\left(\sum_{y} |g(y)|q(y)\right) < \infty.$$

Now

$$E[f(X)g(Y)] = \sum_{x}\sum_{y} f(x)g(y)p(x)q(y)$$

and by the same manipulations as above, this quantity is found to be

$$\left(\sum_{x} f(x)p(x)\right)\left(\sum_{y} g(y)q(y)\right) = E[f(X)]E[g(Y)].$$

E10. Apply the convolution formula (33) to the case $p_k = e^{-\lambda}\lambda^k/k!$, $q_k = e^{-\mu}\mu^k/k!$ to find

$$r_k = e^{-(\lambda+\mu)} \sum_{j=0}^{k} \frac{\lambda^j \mu^{k-j}}{j!(k-j)!}$$

and apply the binomial formula to get

$$\sum_{j=0}^{k} \frac{\lambda^j \mu^{k-j}}{j!(k-j)!} = \frac{1}{k!}\sum_{j=0}^{k} \frac{k!}{j!(k-j)!}\lambda^j\mu^{k-j} = \frac{1}{k!}(\lambda + \mu)^k,$$

and therefore

$$r_k = e^{-(\lambda+\mu)}\frac{(\lambda + \mu)^k}{k!}.$$

E11. $$\sigma^2 = E[X^2] - m^2 = \lambda^2 + \lambda - \lambda^2 = \lambda.$$

E.12.
$$E[X] = \sum_{k=1}^{\infty} kpq^{k-1} = p\sum_{k=1}^{\infty} kq^{k-1} = p\sum_{k=0}^{\infty} \frac{d}{dq}(q^k)$$

$$= p\frac{d}{dq}\sum_{k=0}^{\infty} q^k = p\frac{1}{(1-q)^2} = p\frac{1}{p^2} = \frac{1}{p}.$$

And

$$E[X^2] = \sum_{k=1}^{\infty} k^2 pq^{k-1} = \sum_{k=1}^{\infty} k(k-1)pq^{k-1} + \sum_{k=1}^{\infty} kpq^{k-1}.$$

$$\sum_{k=1}^{\infty} k(k-1)pq^{k-1} = pq\sum_{k=0}^{\infty} \frac{d^2}{dq^2}(q^k) = pq\frac{d^2}{dq^2}\left(\frac{1}{1-q}\right) = pq2\frac{1}{(1-q)^3} = 2\frac{q}{p^2}.$$

Therefore,

$$E[X^2] = 2\frac{q}{p^2} + \frac{1}{p} = \frac{2q+p}{p^2},$$

and by Eq. (38),

$$\sigma^2 = \frac{2q+p}{p^2} - \frac{1}{p^2} = \frac{2q+p-1}{p^2} = \frac{(q+p)+q-1}{p^2} = \frac{1+q-1}{p^2} = \frac{q}{p^2}.$$

E.13. Since $\{X = n\} = \{X \geq n\} - \{X \geq n+1\}$, $P(X = n) = P(X \geq n) - P(X \geq n+1)$. Therefore, $E[X] = \sum_{n=0}^{\infty} nP(X = n) = \sum_{n=0}^{\infty} [nP(X \geq n) - (n+1)P(X \geq n+1) + P(X \geq n+1)] = \sum_{n=0}^{\infty} P(X \geq n+1) = \sum_{n=1}^{\infty} P(X \geq n)$.

E.14. Let $Y = \min(X_1, \ldots, X_m)$. By Eq. (39), $E[Y] = \sum_{n=1}^{\infty} P(Y \geq n)$. But $P(Y \geq n) = P(\min(X_1, \ldots, X_m) \geq n) = P(X_1 \geq n, \ldots, X_m \geq n) = P(X_1 \geq n) \ldots P(X_m \geq n) = r_n^m$.

Case $X_i \sim \mathscr{G}(P)$. $r_n = P(X_i \geq n) = \sum_{k=n}^{\infty} pq^{k-1} = pq^{n-1}(1 + q + \cdots) = pq^{n-1} \times (1/1 - q) = q^{n-1}$, and therefore $\sum_{n=1}^{\infty} r_n^m = \sum_{n=1}^{\infty} (q^m)^{n-1} = 1/(1 - q^m)$. Note that $\min(X_1, \ldots, X_m) \sim \mathscr{G}(1 - q^m)$.

E.15. Let X be any of the X_i's. Then $E[X] = p$ and the variance of X is $E[X^2] - p^2 = p - p^2 = p(1 - p) = pq$. Therefore, $E[S_n] = \sum_{i=1}^{n} E[X_i] = np$, and since the X_i's are independent, the variance of S_n is the sum of the variances of the X_i's, that is, npq.

E.16. The variance of $X_1 + \cdots + X_n$ is $n\sigma^2$ and that of $(X_1 + \cdots + X_n)/n$ is $(1/n^2)(n\sigma^2) = \sigma^2/n$. The standard deviation of $(X_1 + \cdots + X_n)/n$ is therefore $\sigma/\sqrt{n}$.

E.17. Let X_i be the result of the ith experiment: X_i has mean m and variance 10^{-8}, where m is the actual value of the measured quantity. By Eq. (43), taking $\varepsilon = 10^{-4}$,

$$P\left(\left|\frac{X_1 + \cdots + X_n}{n} - m\right| \geq 10^{-4}\right) \leq \frac{10^{-8}}{n \cdot 10^{-8}} = \frac{1}{n}.$$

Therefore, the left-hand side of the above inequality is less than $1 - 0.99 = 10^{-2}$ if $n \geq 100$.

18.
$$E\left[f\left(\frac{S_n}{n}\right)\right] = \sum_{k=0}^{n} f\left(\frac{k}{n}\right)P(S_n = k) = \sum_{k=0}^{n} f\left(\frac{k}{n}\right)\frac{n!}{k!(n-k)!}x^k(1-x)^{n-k}$$

since $S_n \sim B(n, p)$ (see Example 3). The function f is continuous on $[0, 1]$ and therefore uniformly continuous on $[0, 1]$. Therefore to any $\varepsilon > 0$, one can associate a number $\delta(\varepsilon)$ such that if $|y - x| \leqslant \delta(\varepsilon)$, then $|f(x) - f(y)| \leqslant \varepsilon$. Being continuous on $[0, 1]$, f is bounded on $[0, 1]$ by some finite number, say M. Now

$$|P_n(x) - f(x)| = \left| E\left[f\left(\frac{S_n}{n} \right) \right] - f(x) \right|$$

$$= \left| E\left[f\left(\frac{S_n}{n} \right) - f(x) \right] \right| \leqslant E\left[\left| f\left(\frac{S_n}{n} \right) - f(x) \right| \right]$$

$$= E\left[\left| f\left(\frac{S_n}{n} \right) - f(x) \right| 1_A \right] + E\left[\left| f\left(\frac{S_n}{n} \right) - f(x) \right| 1_{\bar{A}} \right]$$

where A is the set of ω's such that $|(S_n(\omega)/n) - x| \leqslant \delta(\varepsilon)$. Since $|f(S_n/n) - f(x)| 1_{\bar{A}} \leqslant 2M 1_{\bar{A}}$, we have

$$E\left[\left| f\left(\frac{S_n}{n} \right) - f(x) \right| 1_{\bar{A}} \right] \leqslant 2M P(\bar{A}) = 2M P\left(\left| \frac{S_n}{n} - x \right| \geqslant \delta(\varepsilon) \right).$$

Also, by definition A and $\delta(\varepsilon)$,

$$E\left[\left| f\left(\frac{S_n}{n} \right) - f(x) \right| 1_A \right] \leqslant \varepsilon.$$

Therefore

$$|P_n(x) - f(x)| \leqslant \varepsilon + 2M P\left(\left| \frac{S_n}{n} - x \right| \geqslant \delta(\varepsilon) \right).$$

But x is the mean of S_n/n, and the variance of S_n/n is $nx(1 - x) \leqslant n/4$. Therefore, by Eq. (43),

$$P\left(\left| \frac{S_n}{n} - x \right| \geqslant \delta(\varepsilon) \right) \leqslant \frac{4}{n[\delta(\varepsilon)]^2}.$$

Finally

$$|f(x) - P_n(x)| \leqslant \varepsilon + \frac{4}{n[\delta(\varepsilon)]^2},$$

and this suffices to prove the convergence in Eq. (46). The convergence is uniform since the right-hand side of the latter inequality does not depend on $x \in [0, 1]$.

E19.
$$g(s) = E[s^X] = \sum_{k=0}^{n} s^k P(X = k) = \sum_{k=0}^{n} s^k \frac{n!}{k!(n - k)!} p^k q^{n-k}$$

$$= \sum_{k=0}^{n} \frac{n!}{k!(n - k)!} (ps)^k q^{n-k} = (ps + q)^n.$$

E20.
$$g(s) = \sum_{k=1}^{\infty} pq^{k-1} s^k = ps \sum_{k=1}^{\infty} (qs)^{k-1}$$

$$= ps \sum_{k=0}^{\infty} (qs)^k = ps/(1 - qs).$$

E21.
$$g(s_1, \ldots, s_k) = \sum_{\substack{m_1=1 \\ m_1+\cdots+m_k=n}}^{n} \cdots \sum_{m_k=1}^{n} \frac{n!}{m_1! \ldots m_k!} p_1^{m_1} \ldots p_k^{m_k} s_1^{m_1} \ldots s_k^{m_k}$$

$$= \sum_{\substack{m_1=1 \\ m_1+\cdots+m_k=n}}^{n} \cdots \sum_{m_k=1}^{n} \frac{n!}{m_1! \ldots m_k!} (p_1 s_1)^{m_1} \ldots (p_k s_k)^{m_k}$$

$$= (p_1 s_1 + \cdots + p_k s_k)^n$$

$$g_1(s_1) = g(s_1, 1, \ldots, 1)$$

$$= (p_1 s_1 + p_2 + \cdots + p_k)^n$$

$$= (p_1 s_1 + (1 - p_1))^n.$$

Therefore, $X_1 \sim \mathscr{B}(n, p_1)$. Similarly $X_i \sim \mathscr{B}(n, p_i)$ for all i, $1 \leqslant i \leqslant n$. Therefore $\sigma_i^2 = np_i q_i$, $\sum \sigma_i^2 = n \sum_{i=1}^{k} (p_i q_i)$. But $X_1 + \cdots + X_k = n$ and therefore has variance 0, being a constant.

E22. $g(s) = (ps + q)^n$, $g'(s) = np(ps + q)^{n-1}$, $g''(s) = n(n-1)p^2(ps + q)^{n-2}$. Therefore, $g'(1) = np(p + q)^{n-1} = np$ and $g''(1) = n(n-1)p^2(p + q)^{n-2} = n(n-1)p^2$. By Eq. (56) $m = E[X] = np$. And by Eq. (57), $\sigma^2 = n(n-1)p^2 + np - n^2 p^2 = np(1 - p) = npq$.

E23. $g(s) = e^{\lambda(s-1)}$, $g'(s) = \lambda e^{\lambda(s-1)}$, $g''(s) = \lambda^2 e^{\lambda(s-1)}$. Therefore, $g'(1) = \lambda$, $g''(1) = \lambda^2$. By Eq. (56), $m = E[X] = \lambda$. By Eq. (57), $\sigma^2 = \lambda^2 + \lambda - \lambda^2 = \lambda$.

E24. $g_1(s) = E[s^{X_1}] = e^{\lambda_1(s-1)}$, $g_2(s) = E[s^{X_2}] = e^{\lambda_2(s-1)}$ and $g(s) = E[s^{X_1+X_2}] = E[s^{X_1}s^{X_2}] = E[s^{X_1}] \cdot E[s^{X_2}] = e^{(\lambda_1+\lambda_2)(s-1)}$. Therefore, since the generating function characterizes the distribution, $X_1 + X_2 \sim P(\lambda_1 + \lambda_2)$.

E25. By Eq. (57), $E[Y] = (g_T \circ g_X)'(1) = g_T'(g_X(1))g_X'(1)$. Also by Eq. (57), $g_X'(1) = E[X_1]$ and $g_T'(1) = E[T]$. Now $g_X(1) = 1$ and therefore $E[Y] = g_T'(1)g_X'(1) = E[T]E[X_1]$.

E26. The equality $\phi_{n+1}(s) = \phi_n(g_z(s))$ is just a special case of Eq. (60a). Iteration of this relation yields

$$\phi_{n+1}(s) = \phi_0(\underbrace{g_z(g_z(\ldots(g_z(s))\ldots)))}_{n+1 \text{ times}}.$$

When $X_0 = 1$, $\phi_0(s) = E[s^{X_0}] = E[s'] = s$, and therefore

$$\phi_{n+1}(s) = \underbrace{g_z(g_z(\ldots(g_z(s))\ldots))}_{n+1 \text{ times}}, \qquad \text{qed.}$$

We now use Eqs. (56) and (57), which gives $m_n = \phi_n'(1) = g_z'[\phi_{n-1}(1)]\phi_{n-1}'(1) = g_z'(1) \cdot \phi_{n-1}'(1) = m \cdot m_{n-1}$ $(n \geqslant 1)$. If $X_0 = 1$ then $m_0 = 1$, so that $m_n = m^n$ $(n \geqslant 0)$. Also $v_n = \phi_n''(1) + m_n - m_n^2$, $\sigma^2 = g_z''(1) + m - m^2$. But $\phi_n''(1) = g_z''[\phi_{n-1}(1)]$ $\phi_{n-1}'(1)^2 + g_z'[\phi_{n-1}(1)]\phi_{n-1}''(1) = g_z''(1)\phi_{n-1}'(1)^2 + g_z'(1)\phi_{n-1}''(1)$, therefore $v_n = \sigma^2 m^{2n-2} + mv_{n-1}$ $(n \geqslant 1)$. If $X_0 = 1$ then $v_0 = \sigma_0^2 = 0$, therefore, for $n \geqslant 0$,

$$v_n = \begin{cases} \sigma^2 m^{n-1} \dfrac{1-m^n}{1-m} & \text{if } m \neq 1 \\[2ex] n\sigma^2 & \text{if } m = 1. \end{cases}$$

If $X_0 = k$, we have k independent branching processes so that

$$\begin{cases} m_n = km^n \\[1ex] v_n = \begin{cases} k^2 \sigma^2 m^{n-1} \dfrac{1-m^n}{1-m} & \text{if } m \neq 1 \\[2ex] k^2 n\sigma^2 & \text{if } m = 1. \end{cases} \end{cases}$$

E27. If $X_0 = k$, extinction occurs when the k independent branching processes become extinct. Therefore,

$$P_e^{(k)} = (P_e^{(1)})^k.$$

E28. $H(Y) = -E[\log p(Y)] = -\sum_{y \in F} p(y) \log p(y)$. Therefore,

$$H(X_1) = -\sum_{i=1}^{k} p_i \log p_i.$$

With obvious notations,

$$H(X_1, \ldots, X_n) = -E[\log p(X_1, \ldots, X_n)] = -E[\log p(X_1) \ldots p(X_n)]$$

$$= -E[\log p(X_1) + \cdots + \log p(X_n)] = -\sum_{j=1}^{n} E[\log p(X_j)]$$

$$= \sum_{j=1}^{n} H(X_j) = nH(X_1)$$

since $H(X_j) = H(X_1)$ for all j.

E29. Assume without loss of generality that $p_i > 0$ $(1 \leqslant i \leqslant k)$. Since $\log q_i/p_i \leqslant (q_i/p_i) - 1$ with equality if and only if $p_i = q_i$, we have

$$\sum_{i=1}^{k} p_i \log \frac{q_i}{p_i} \leqslant \sum_{i=1}^{k} p_i \left(\frac{q_i}{p_i} - 1 \right) = \sum_{i=1}^{n} p_i - \sum_{i=1}^{n} p_i = 0$$

with equality if and only if $p_i = q_i$ $(1 \leqslant i \leqslant k)$. Hence, Gibbs' inequality. Take $q_i = 1/k$ $(1 \leqslant i \leqslant k)$ in Gibbs' inequality to obtain

$$H(p_1, \ldots, p_k) \leqslant \log k.$$

The equality holds if and only if $p_i = 1/k$ $(1 \leqslant i \leqslant k)$. Clearly $H(X_1) \geqslant 0$. It is equal to 0 if and only if $p_i \log p_i = 0$ $(1 \leqslant i \leqslant k)$, i.e., $p_i = 0$ or 1 $(1 \leqslant i \leqslant k)$. Now $\sum_{i=1}^{k} p_i = 1$ and therefore there exists one and only one j such that $p_j = 1$.

E30. Look at the subtree with $\textcircled{a_i}$ as a root, and look at the 2^m nodes of the binary tree at level m where $m = \sup_{1 \leqslant i \leqslant k} l_i(c)$. (See figure below for an explanation of the terms.)

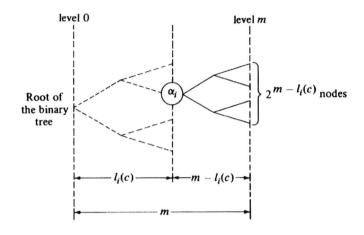

Since c is a prefix code, there is no other α_j on the subtree originating from α_i. Therefore, clearly $2^m \geq \sum_{i=1}^{k} 2^{m-l_i(c)}$.

Suppose $l_1 \leq l_2 \leq \cdots \leq l_k = m$. Look at the nodes at level m. Starting from the top, reserve 2^{m-l_1} nodes for α_1, then the next 2^{m-l_2} nodes for α_2, etc. There are enough nodes since $\sum_{i=1}^{k} 2^{m-l_i} \leq 2^m$. Now obtain the code word for α_i as the following figure shows.

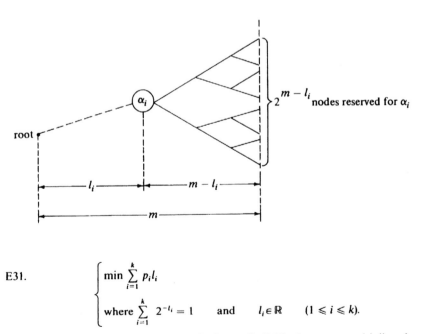

E31.

$$\begin{cases} \min \sum_{i=1}^{k} p_i l_i \\ \text{where } \sum_{i=1}^{k} 2^{-l_i} = 1 \quad \text{and} \quad l_i \in \mathbb{R} \quad (1 \leq i \leq k). \end{cases}$$

(Why can we take equality in Kraft's inequality?) Use Lagrange multipliers, i.e., define

$$f(\lambda, l_1, \ldots, l_k) = \sum_{i=1}^{k} p_i l_i + \lambda \left(\sum_{i=1}^{k} e^{-l_i \log 2} - 1 \right),$$

(where log is natural log), and set

$$\begin{cases} \dfrac{\partial f}{l_i} = 0 & (1 \leqslant i \leqslant k) \\[2mm] \dfrac{\partial f}{\partial \lambda} = 0 \end{cases}$$

i.e.,

$$\begin{cases} p_i - \lambda \log 2 \cdot 2^{-l_i} = 0 & (1 \leqslant i \leqslant k) \\[2mm] \displaystyle\sum_{i=1}^{k} 2^{-l_i} = 1. \end{cases}$$

The solution is $l_i = -\log_2 p_i$ $(1 \leqslant i \leqslant k)$ and $\lambda = 1/\log 2$. But these l_i's are in general not admissible since they are not in general integers. Therefore, take instead l_i equal to the smallest integer larger than $-\log_2 p_i$. Then clearly

$$-\sum_{i=1}^{k} p_i \log_2 p_i \leqslant \sum_{i=1}^{k} p_i l_i \leqslant \sum_{i=1}^{k} p_i (-\log_2 p_i + 1) = -\sum_{i=1}^{k} p_i \log_2 p_i + 1.$$

E32. The first part is easy: simply interchange nodes in the tree. As for Huffman's procedure, an example will suffice:

$$p = (p_1, \ldots, p_8) = (0.05, 0.05, 0.05, 0.1, 0.1, 0.2, 0.2, 0.25).$$

Take the two smallest probabilities (in our example, we may choose p_1, p_2 or p_2, p_3). For graphical reasons, it is better to choose the first two probabilities in the lexicographical order. Do this:

$p_1 + p_2 = 0.1$

$p_1 = 0.05$ $p_2 = 0.05$ $p_3 = 0.05$ $p_4 = 0.1$ $p_5 = 0.1$ $p_6 = 0.2$ $p_7 = 0.2$ $p_8 = 0.25$

Now select among $(p_1 + p_2, p_3, \ldots, p_8)$ the two smallest numbers. Here $p_1 + p_2$, p_3. Do this:

$p_1 + p_2 + p_3 = 0.15$

0.05   0.05   0.05   0.1   0.1   0.2   0.2   0.25

And the procedure continues

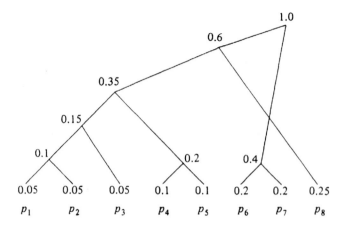

Make this graph look more like a decent tree, and call 0 a branch going to the left, 1 a branch going to the right:

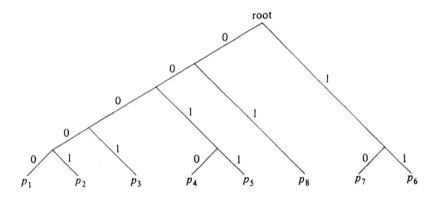

The code for α_i is obtained by reading the branches when going from the root to p_i.

$$p_1 : 0\,0\,0\,0\,0$$
$$p_2 : 0\,0\,0\,0\,1$$
$$p_3 : 0\,0\,0\,1$$
$$p_4 : 0\,0\,1\,0$$
$$p_5 : 0\,0\,1\,1$$
$$p_6 : 1\,1$$
$$p_7 : 1\,0$$
$$p_8 : 0\,1.$$

E33. $H_2(X_1,\ldots,X_n) \leqslant L(c^{(n)}) \leqslant H_2(X_1,\ldots,X_n) + 1$. But $H_2(X_1,\ldots,X_n) = nH_2(X_1)$, therefore

$$H_2(X_1) \leqslant \frac{L(c^{(n)})}{n} \leqslant H_2(X_1) + \frac{1}{n}.$$

E34. It suffices to encode the source by blocks of three. The statistics of 3-blocks are easily computed. After rearrangement in the order of increasing probability,

$$
\begin{array}{ll}
0\,0\,0 & p_0 = 1/64 \\
0\,0\,1 & p_1 = 3/64 \\
0\,1\,0 & p_2 = 3/64 \\
1\,0\,0 & p_3 = 3/64 \\
0\,1\,1 & p_4 = 9/64 \\
1\,0\,1 & p_5 = 9/64 \\
1\,1\,0 & p_6 = 9/64 \\
1\,1\,1 & p_7 = 27/64.
\end{array}
$$

A Huffman code for 3-blocks is constructed:

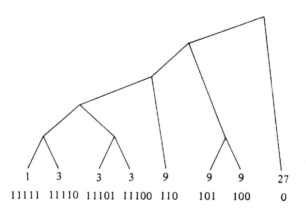

| 1 | 3 | 3 | 3 | 9 | 9 | 9 | 27 |
| 11111 | 11110 | 11101 | 11100 | 110 | 101 | 100 | 0 |

The average length per symbol is

$$\tfrac{1}{3}\cdot\tfrac{1}{64}(1 \times 5 + 3 \times 5 + 3 \times 5 + 3 \times 5 + 9 \times 3 + 9 \times 3 + 9 \times 3 + 27 \times 1)$$
$$= \tfrac{158}{192} \simeq 0.822.$$

(With blocks of length 2, the average length per symbol after Huffman encoding is $27/32 \simeq 0.843$.)

E35. Place the coins in line, and name seven configurations: configuration 7 has no bad coin. For $1 \leqslant i \leqslant 6$, i is the configuration for which the bad coin is in position i. The probability for configuration 7 is $\tfrac{2}{3}$. For each configuration i, $1 \leqslant i \leqslant 6$, the probability is $\frac{1}{3} \cdot \frac{1}{6}$. The Huffman code for this set of probabilities is shown in the figure below.

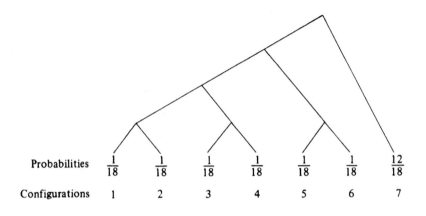

Probabilities $\quad \frac{1}{18} \qquad \frac{1}{18} \qquad \frac{1}{18} \qquad \frac{1}{18} \qquad \frac{1}{18} \qquad \frac{1}{18} \qquad \frac{12}{18}$

Configurations $\quad$ 1 $\qquad$ 2 $\qquad$ 3 $\qquad$ 4 $\qquad$ 5 $\qquad$ 6 $\qquad$ 7

To discriminate between 7 and 1, 2, 3, 4, 5, or 6, weigh the first three coins against the last three coins. The rest of the procedure is easy.

Remark. With seven bags, the Huffman search procedure is not implementable with a scale.

CHAPTER 3

Probability Densities

1. Expectation of Random Variables with a Density

1.1. Univariate Probability Densities

Let X be a real random variable defined on (Ω, F, P) with the cumulative distribution function (c.d.f.) (see Chapter 1, Section 3.2)

$$\boxed{F(x) = P(X \leqslant x)}.$$ (1)

If there exists a non-negative function f such that

$$\boxed{F(x) = \int_{-\infty}^{x} f(y)\,dy}\,,$$ (2)

then f is called the *probability density* (p.d.) of X. Recall (Chapter 1, Section 3.2) that

$$\int_{-\infty}^{+\infty} f(y)\,dy = 1.$$ (3)

The following are a few classic probability densities.

The Uniform Density on $[a, b]$. This is certainly the simplest probability density (Fig. 1).

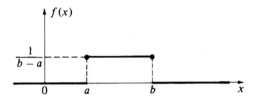

Figure 1. The uniform probability density on $[a, b]$.

$$f(x) = \begin{cases} \dfrac{1}{b-a} & \text{if } x \in [a,b] \\ 0 & \text{otherwise.} \end{cases} \tag{4}$$

A random variable X with the probability density of Eq. (4) is said to be *uniformly distributed* on $[a, b]$. This is denoted by

$$X \sim \mathcal{U}([a,b]).$$

The Exponential Density of Parameter $\lambda > 0$ (Fig. 2).

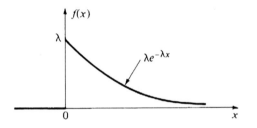

Figure 2. The exponential p.d.

$$f(x) = \begin{cases} \lambda e^{-\lambda x} & \text{if } x \geq 0 \\ 0 & \text{otherwise.} \end{cases} \tag{5}$$

The corresponding cumulative distribution function is (Fig. 3):

$$F(x) = \begin{cases} 1 - e^{-\lambda x} & \text{if } x \geq 0 \\ 0 & \text{otherwise.} \end{cases} \tag{6}$$

A random variable with the probability density of Eq. (5) is an *exponential random variable*. This is denoted by

$$X \sim \mathscr{E}(\lambda).$$

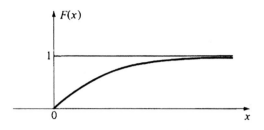

Figure 3. The exponential c.d.f.

The Gaussian Density. This is the most famous probability density. It is defined for $m \in \mathbb{R}$, $\sigma \in \mathbb{R}_+ - \{0\}$ (Fig. 4) by

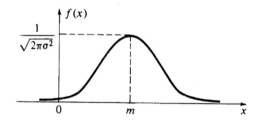

Figure 4. The Gaussian p.d.

$$f(x) = \frac{1}{\sigma\sqrt{2\pi}} e^{-(1/2)[(x-m)/\sigma]^2} .$$ (7)

We shall soon give the probabilistic interpretation of the parameters $m \in \mathbb{R}$ and $\sigma^2 \in \mathbb{R}_+$. If X admits such density, then we note

$$X \sim \mathcal{N}(m, \sigma^2),$$

and say that X is a *Gaussian random variable.* When $X \sim \mathcal{N}(0, 1)$ we say that X is a *standard Gaussian random variable.*

E1 Exercise. Verify that the function f of Eq. (7) satisfies Eq. (3).

E2 Exercise. Show that if X is a standard Gaussian random variable and $\sigma > 0$, then $Y = \sigma X + m \sim \mathcal{N}(m, \sigma^2)$. Conversely, show that if $Y \sim \mathcal{N}(m, \sigma^2)$, then $X = (Y - m)/\sigma \sim \mathcal{N}(0, 1)$.

The Gamma Density. Let α and β be two strictly positive real numbers and define

$$f(x) = \begin{cases} \dfrac{\beta^\alpha}{\Gamma(\alpha)} x^{\alpha-1} e^{-\beta x} & \text{if } x > 0 \\ \\ 0 & \text{otherwise} \end{cases} \tag{8}$$

where the *gamma function* Γ is defined by

$$\Gamma(\alpha) = \int_0^\infty u^{\alpha-1} e^{-u} \, du \,. \tag{9}$$

Integration by parts yields the functional equation

$$\Gamma(\alpha) = (\alpha - 1)\Gamma(\alpha - 1) \qquad (\alpha > 1). \tag{10}$$

Since $\Gamma(1) = 1$, it follows that, if n is a strictly positive integer,

$$\Gamma(n) = (n - 1)! \tag{11}$$

E3 Exercise. Check that f is a probability density.

Density (8) is called the *gamma probability density* of parameters α and β. If X admits such density, we note

$$X \sim \gamma(\alpha, \beta),$$

and say that X is a *gamma distributed random variable* (Fig. 5).

When $\alpha = 1$, the gamma distribution is simply the exponential distribution

$$\gamma(1, \beta) \equiv \mathscr{E}(\beta).$$

When $\alpha = n/2$ and $\beta = \frac{1}{2}$, the corresponding distribution is called the *chi-square distribution with n degrees of freedom*. When X admits this density, this is denoted by

$$X \sim \chi_n^2.$$

Expectation. Let X be a random variable with the probability density f. Let g be a function from $\mathbb{R}$ into $\mathbb{R}$ for which the quantity $\int_{-\infty}^{+\infty} g(x) f(x) \, dx$ has a meaning (as a Riemann integral, for instance). Then such a quantity is called the (mathematical) expectation of $g(X)$ and is denoted by $E[g(X)]$.

1.2. Mean and Variance

When they are defined, the quantities

$$m = E[X] = \int_{-\infty}^{+\infty} x f(x) \, dx \tag{12}$$

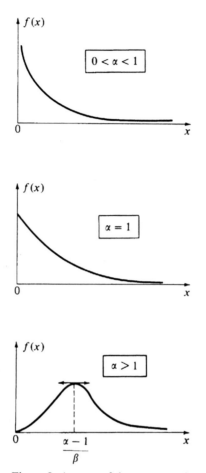

Figure 5. Aspects of the gamma p.d.

and

$$\sigma^2 = E[(X - m)^2] = \int_{-\infty}^{+\infty} (x - m)^2 f(x)\, dx \qquad\qquad (13)$$

are called, respectively, the *mean* and the *variance* of X. The *standard deviation* of X is σ, the nonnegative square root of the variance. By the linearity of the integral

$$\int_{-\infty}^{+\infty} (x - m)^2 f(x)\, dx = \int_{-\infty}^{+\infty} x^2 f(x)\, dx - 2m \int_{-\infty}^{+\infty} x f(x)\, dx + m^2 \int_{-\infty}^{+\infty} f(x)\, dx$$

and therefore, in view of Eqs. (13) and (3)

$$\boxed{\sigma^2 = E[X^2] - m^2}. \tag{14}$$

Also, again using the linearity of the integral,

$$\int_{-\infty}^{+\infty} (ax + b)f(x)\,dx = a \int_{-\infty}^{+\infty} xf(x)\,dx + b \int_{-\infty}^{+\infty} f(x)\,dx,$$

so that by Eq. (3),

$$E[aX + b] = aE[X] + b,$$

that is, if one adopts the notation $m_Y = E[Y]$,

$$\boxed{m_{aX+b} = am_X + b}. \tag{15}$$

With a similar notation for the variance, one has

$$\boxed{\sigma^2_{aX+b} = a^2 \sigma^2_X}. \tag{16}$$

Indeed, $\quad \sigma^2_{aX+b} = E\{[(aX + b) - m_{aX+b}]^2\} = E[(aX + b - am_X - b)^2] = E[a^2(X - m_X)^2] = a^2 E[(X - m_X)^2] = a^2 \sigma^2_X.$

E4 Exercise. Show that if $X \sim \gamma(\alpha, \beta)$, then $m_X = \alpha/\beta$ and $\sigma^2_X = \alpha/\beta^2$.

E5 Exercise. Compute the mean and variance of X when $X \sim \mathcal{U}([a, b])$, $X \sim \mathcal{E}(\lambda)$ and $X \sim \mathcal{N}(0, 1)$.

From the solution of the above exercise we shall extract

$$\boxed{m_X = \frac{1}{\lambda} \quad [X \sim \mathcal{E}(\lambda)]} \tag{17}$$

and

$$\boxed{m_X = 0,\ \sigma^2_X = 1 \quad [X \sim \mathcal{N}(0, 1)]}. \tag{18}$$

From Exercise E2, we know that if $X \sim \mathcal{N}(m, \sigma^2)$, then $X = \sigma Y + m$ where $Y \in \mathcal{N}(0, 1)$. Therefore, from Eqs. (15), (16), and (18) follows the interpretation of the parameters m and σ^2 in the $\mathcal{N}(m, \sigma^2)$ probability density of Eq. (7):

$$m_X = m,\ \sigma^2_X = \sigma^2 \quad [X \sim \mathcal{N}(m, \sigma^2)]. \tag{19}$$

A Case Where the Mean and Variance Do Not Exist. The nonnegative function

$$f(x) = \frac{1}{\pi} \frac{1}{1 + x^2} \qquad (x \in \mathbb{R}) \tag{20}$$

is such that $\int_{-\infty}^{+\infty} f(x)\,dx = 1$. It is therefore a probability density. The quantity $\int_{-\infty}^{+\infty} xf(x)\,dx$ is not defined as a Riemann integral since it leads to the un-defined form $+\infty - \infty$. Indeed, by definition of a proper Riemann integral, $\int_{-\infty}^{+\infty} xf(x)\,dx = \lim_{A\downarrow-\infty,B\uparrow+\infty} \int_A^B xf(x)\,dx = \lim_{A\downarrow-\infty} \int_A^0 xf(x)\,dx + \lim_{B\uparrow\infty} \cdot \int_0^B xf(x)\,dx$. Note, however, that the extended Riemann integral, defined to be equal to $\lim_{N\uparrow\infty} \int_{-N}^{+N} xf(x)\,dx$, exists and equals 0. Although in view of the symmetry of f (see Fig. 6) it is natural to define its mean to be 0, the convention among probabilists is that the mean does not exist in this case because of the nonexistence of $\int_{-\infty}^{+\infty} xf(x)\,dx$ as a proper Riemann integral. Of course, if m is not defined, σ^2 is not defined.

The function f in Eq. (20) is called the *Cauchy probability density* (Fig. 6), and any random variable with such a probability density is a *Cauchy random variable*.

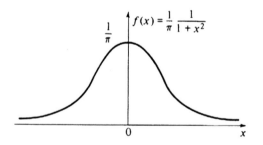

Figure 6. The Cauchy p.d.

One could find it exaggerated to insist that a Cauchy random variable has no mean when it is graphically obvious that it has one. However, it is not only illicit but also dangerous to maintain that a Cauchy random variable has mean 0, for the following reason. In Chapter 5, the strong law of large numbers states that if $(X_n, n \geq 1)$ is a sequence of independent and identically distributed random variables with mean m, the empirical frequency $(X_1 + \cdots + X_n)/n$ tends "almost-surely" (this phrase will be explained in Chapter 5) towards m. At the end of Chapter 5, there is a remark concerning the case where the X_n's are Cauchy random variables and it is shown that in this case, the em-pirical frequency does not tend to 0. This of course does not imply that the strong law of large numbers is false, but only that it does not apply to Cauchy random variables. Indeed the strong law of large numbers con-siders random variables X_n *with a mean*, and this implies, by definition, that $E|X_n| < \infty$.

1.3. Chebyshev's Inequality

Basic Properties of Expectation. We will take as intuitively obvious the following fact. If the probability of event $\{\omega | X(\omega) \leqslant Y(\omega)\}$ is one, which is written $X \leqslant Y$ P-as, then $E[X] \leqslant E[Y]$. This is the *monotonicity* property of expectation. It is true in all situations where $E[X]$ and $E[Y]$ are defined. Also the *linearity* property $E[aX + bY] = aE[X] + bE[Y]$ holds whenever the expectations in both sides of the equality have a meaning.

At the end of this chapter, we sketch the general theory of expectation for random variables that are not necessarily discrete or that do not admit a probability density, but for now, we cannot produce a rigorous proof, even in the particular case of the existence of a probability density.

Another basic property of expectation is the following one: $|E[X]| \leqslant E[|X|]$. In the case where X admits a density, this is simply

$$\left| \int_{-\infty}^{+\infty} xf(x)\,dx \right| = \left| \int_{-\infty}^{0} xf(x)\,dx + \int_{0}^{\infty} xf(x)\,dx \right|$$

$$= \left| -\int_{-\infty}^{0} |x| f(x)\,dx + \int_{0}^{\infty} |x| f(x)\,dx \right|$$

$$\leqslant \int_{-\infty}^{+\infty} |x| f(x)\,dx.$$

In view of future reference, the above properties will be recapitulated:

$$X \leqslant Y, \text{ P-as} \Rightarrow E[X] \leqslant E[Y]. \tag{21}$$

$$E[aX + bY] = aE[X] + bE[Y]. \tag{22}$$

$$|E[X]| \leqslant E[|X|]. \tag{23}$$

These elementary properties will now be used to derive two famous inequalities already encountered in Chapter 2 for discrete random variables.

Markov's Inequality. Let X be a random variable and let f be a function from $\mathbb{R}$ into $\mathbb{R}_+$. Then

$$\boxed{P[f(X) \geqslant a] \leqslant \frac{E[f(X)]}{a}} \qquad (a > 0). \tag{24}$$

PROOF OF EQ. (24). Observe that for all $x \in \mathbb{R}$,

$$a1_{f \geqslant a}(x) \leqslant f(x) \qquad (x \in \mathbb{R})$$

where $1_{f \geqslant a}(x)$ is equal to 1 if $f(x) \geqslant a$ and to 0 otherwise, i.e., $1_{f \geqslant a}$ is the indicator function of the set $C = \{x | f(x) \geqslant a\}$ (Fig. 7).

From the above inequality, we deduce the inequality between random

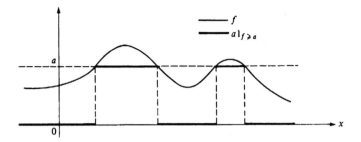

Figure 7. The function $a1_{f \geq a}$.

variables

$$a \cdot 1_C(X(\omega)) \leq f[X(\omega)] \qquad (\omega \in \Omega).$$

Since $1_C(X(\omega)) = 1_A(\omega)$ where $A = \{\omega | f(X(\omega)) \geq a\}$, the last inequality reads

$$a \cdot 1_A \leq f(X)$$

(where the ω dependency is no longer explicit). By the monotonicity property (21),

$$E[a1_A] \leq E[f(X)].$$

By the linearity property $E[a1_A] = aE[1_A]$, and since $E[1_A] = P(A) = P(f(X) \geq a)$, Eq. (24) follows. □

Chebyshev's inequality applies to random variables X for which the mean m and variance σ^2 are defined. It is a special case of Markov's inequality with $f(x) = (x - m)^2$ and $a = \varepsilon^2$ where $\varepsilon > 0$:

$$\boxed{P(|X - m| \geq \varepsilon) \leq \frac{\sigma^2}{\varepsilon^2} \qquad (\varepsilon > 0).} \tag{25}$$

Remark. In the proof of Markov–Chebyshev's inequalities, only properties (21) and (22) have been used. Since such properties are true for any type of random variables (discrete, with a density, etc.), inequalities (24) and (25) also hold in the general case (see Section 4). Note also that Eqs. (14), (15), and (16) also hold in the general case since the only property of expectation used in proving them was linearity.

Null Variance and Almost Constant Random Variables. A random variable X is said to be P-almost surely null (notation: $X = 0$, P-as) if and only if the event $\{\omega | X(\omega) = 0\}$ has probability 1. Now

$$\{\omega \,||X(\omega)| > 1\} = \bigcup_{n=1}^{\infty} \left\{\omega \,||X(\omega)| \geqslant \frac{1}{n}\right\}$$

since for $|X|$ to be strictly positive, it is necessary and sufficient that $|X|$ be larger than $1/n$ for *some* $n \geqslant 1$. In more concise notation,

$$\{|X| > 0\} = \bigcup_{n=1}^{\infty} \left\{|X| \geqslant \frac{1}{n}\right\}. \tag{26}$$

By the sub-σ-additivity property of probability [Eq. (6) of Chapter 1],

$$P(|X| > 0) \leqslant \sum_{n=1}^{\infty} P\left(|X| \geqslant \frac{1}{n}\right). \tag{27}$$

Let now $X = Y - m_Y$ where Y is a random variable with *zero variance*: $\sigma_Y^2 = 0$. By Chebyshev's inequality

$$P\left(|X| \geqslant \frac{1}{n}\right) = P\left(|Y - m_Y| \geqslant \frac{1}{n}\right) \leqslant n^2 \sigma_Y^2 = 0.$$

Therefore, from Eq. (27), $P(|Y - m_Y| > 0) = 0$, i.e., $P(|Y - m_Y| = 0) = 1$. In summary,

$$\boxed{\sigma_Y^2 = 0 \Rightarrow Y = m_Y, \quad \text{P-as}} \tag{28}$$

A random variable Y with null variance is therefore P-as equal to a deterministic constant. One can also say that Y is almost-surely constant.

1.4. Characteristic Function of a Random Variable

The definition of expectation is readily extended to the case where X is a *complex random variable*, that is,

$$X = X_1 + iX_2, \tag{29}$$

where X_1 and X_2 are real random variables. The symbol $E[X]$ is defined, if and only if both $E[X_1]$ and $E[X_2]$ are defined and finite, by

$$E[X] = E[X_1] + iE[X_2]. \tag{30}$$

Properties (22) and (23) remain valid when X, Y, a, and b are complex. Therefore, for any real random variable X with a probability density f_X, the characteristic function $\phi_X : \mathbb{R} \to \mathbb{R}$ given by

$$\boxed{\phi_X(u) = E[e^{iuX}] = \int_{-\infty}^{+\infty} e^{iux} f_X(x)\, dx} \tag{31}$$

is defined since $E[\cos(uX)]$ and $E[\sin(uX)]$ are finite. Indeed, for instance, $|E[\cos(uX)]| \leqslant E[|\cos(uX)|] \leqslant E[1] = 1$.

E6 Exercise. Show that

$$\phi_X(u) = \frac{\lambda}{\lambda - iu} \qquad [X \sim \mathscr{E}(\lambda)] \tag{32}$$

and

$$\phi_X(u) = e^{-u^2/2} \qquad [X \sim \mathscr{N}(0, 1)] \tag{33}$$

E7 Exercise. Show that

$$\phi_X(u) = \left(1 - i\frac{u}{\beta}\right)^{-\alpha} \qquad [X \sim \gamma(\alpha, \beta)] \tag{34}$$

Let X be a real random variable with the characteristic function ϕ_X. Then, for any real numbers a and b, the characteristic function ϕ_{aX+b} of $aX + b$ is given by

$$\phi_{aX+b}(u) = e^{iub}\phi_X(au) \tag{35}$$

Indeed,

$$\phi_{aX+b}(u) = E[e^{iu(aX+b)}] = E[e^{iuaX}e^{iub}] = e^{iub}E[e^{i(au)X}].$$

E8 Exercise. Show that

$$\phi_X(u) = e^{ium - (1/2)\sigma^2 u^2} \qquad [X \sim \mathscr{N}(m, \sigma^2)] \tag{36}$$

Remark. Equation (35) is true in general, as one can check by looking at the proof where only the linearity property of expectation was used.

The characteristic function ϕ_X of the real random variable X admitting the probability density f_X is the Fourier transform of f_X. From the theory of Fourier transforms, we know that if

$$\int_{-\infty}^{+\infty} |\phi_X(u)| \, du < \infty, \tag{*}$$

the inverse formula

$$f_X(x) = \frac{1}{2\pi} \int_{-\infty}^{+\infty} e^{-iux} \phi_X(u) \, du$$

holds. A rigorous statement would be as follows: the two functions $x \to f_X(x)$ and $x \to (1/2\pi)\int_{-\infty}^{+\infty} e^{-iux}\phi_X(u)\,du = h(x)$ are "almost equal" in the sense that for any function $g : \mathbb{R} \to \mathbb{R}$, $\int_{-\infty}^{+\infty} g(x)f_X(x)\,dx$ and $\int_{-\infty}^{+\infty} g(x)h(x)\,dx$ are both defined and are equal when either one is defined.

Suppose now that $\phi(u)$ is known to be the characteristic function of two random variables X and Y with a probability density. Then $P(X \leqslant x) = P(Y \leqslant y)$. Indeed, from what we just stated, $\int_{-\infty}^{x} f_X(z)\,dz = \int_{-\infty}^{x} f_Y(z)\,dz$. It turns out (and we will admit this) that requirement (∗) and the existence of a probability density are superfluous. The general statement is: Let X and Y be two real random variables such that

$$E[e^{iuX}] = E[e^{iuY}] \qquad (\forall u \in \mathbb{R}) \tag{37}$$

then

$$P(X \leqslant x) = P(Y \leqslant y) \qquad (\forall x \in \mathbb{R}) \tag{38}$$

and [this is in fact equivalent to Eq. (38)]

$$E[g(X)] = E[g(Y)] \tag{39}$$

for any function g for which $E[g(X)]$ is defined.

One can also say that the characteristic function uniquely determines the law of a random variable.

E9 Exercise. Compute the Fourier transform of $e^{-a|x|}$ when $a > 0$. Deduce from the result that the characteristic function of the Cauchy distribution is $e^{-|u|}$.

Remark. It should be emphasized at this point that two random variables with the same distribution function are not necessarily identical random variables. For instance, take $X \sim \mathcal{N}(0, 1)$ and $Y = -X$. Then $Y \sim \mathcal{N}(0, 1)$, since $P(Y \leqslant y) = P(-X \leqslant y) = P(X \geqslant -y)$ and $P(X \geqslant -y) = P(X \leqslant y)$ whenever X admits a density f_X that is an even function, i.e., $f_X(x) = f_X(-x)$ for all $x \in \mathbb{R}$.

2. Expectation of Functionals of Random Vectors

2.1. Multivariate Probability Densities

Let $X_1, \ldots, X_n$ be real valued random variables. The vector $X = (X_1, \ldots, X_n)$ is then called a real random vector of dimension n. The function $F_X : \mathbb{R}^n \to \mathbb{R}_+$

defined by

$$F_X(x_1,\ldots,x_n) = P(X_1 \leqslant x_1,\ldots,X_n \leqslant x_n) \tag{40}$$

is the cumulative distribution function of X. If

$$F_X(x_1,\ldots,x_n) = \int_{-\infty}^{x_1} \cdots \int_{-\infty}^{x_n} f_X(y_1,\ldots,y_n)\,dy_1 \ldots dy_n \tag{41}$$

for some nonnegative function $f_X : \mathbb{R}^n \to \mathbb{R}$, f_X is called the probability density of X. It satisfies

$$\int_{-\infty}^{+\infty} \cdots \int_{-\infty}^{+\infty} f_X(x_1,\ldots,x_n)\,dx_1 \ldots dx_n = 1. \tag{42}$$

It is sometimes convenient to use more concise notation such as $F_X(x) = P(X \leqslant x)$ instead of Eq. (40), or $\int_{\mathbb{R}^n} f_X(x)\,dx = 1$ instead of Eq. (42). On the contrary, the notation can be more detailed. For instance, one can write $F_{X_1,\ldots,X_n}$ instead of F_X, or $f_{X_1,\ldots,X_n}$ instead of f_X.

E10 Exercise. Recall the probabilistic model of Examples 1 and 3 of Chapter 1, where $\Omega = [0,1]^2$, $P(A) = S(A)$ (area of A) and X and Y are defined in Fig. 8. What is the probability density of the two-dimensional random vector $Z = (X, Y)$?

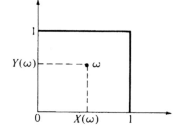

Figure 8. A random point in the unit square.

Marginal Density. Let $X = (X_1,\ldots,X_n)$ be an n-dimensional random vector with a probability density f_X. Let l and m be two integers such that $l + m = n$, and define

$$Y = (X_1,\ldots,X_l), \qquad Z = (X_{l+1},\ldots,X_{l+m}).$$

The l-dimensional random vector Y admits the distribution

$$F_Y(y_1,\ldots,y_l) = P(Y_1 \leqslant y_1,\ldots,Y_l \leqslant y_l) = P(X_1 \leqslant y_1,\ldots,X_l \leqslant y_l),$$

and therefore

$$F_Y(y_1,\ldots,y_l) = \int_{-\infty}^{y_1} \cdots \int_{-\infty}^{y_l} \int_{-\infty}^{+\infty} \cdots \int_{-\infty}^{+\infty} f_X(t_1,\ldots,t_l,t_{l+1},\ldots,t_n)\,dt_1 \ldots dt_n.$$

In view of Fubini's theorem, the above equality can be written

$$F_Y(y_1,\ldots,y_l)$$

$$= \int_{-\infty}^{y_1} \cdots \int_{-\infty}^{y_l} \left[\int_{-\infty}^{+\infty} \cdots \int_{-\infty}^{+\infty} f_X(t_1,\ldots,t_l,t_{l+1},\ldots,t_n)\,dt_{l+1}\ldots dt_n \right] dt_1 \ldots dt_l.$$

From this it is clear that the probability density f_Y of $Y = (X_1,\ldots,X_l)$ is

$$f_Y(y_1,\ldots,y_l) = \int_{-\infty}^{+\infty} \cdots \int_{-\infty}^{+\infty} f_X(y_1,\ldots,y_l,t_{l+1},\ldots,t_n)\,dt_{l+1}\ldots dt_n \,. \qquad (43)$$

The density f_Y is called the *marginal density* of Y.

Remark. The qualification "marginal" is not very meaningful, and one might as well call f_Y the density of Y. It is in fact Y that is a "marginal" of X, i.e., a "subvector" of X.

E11 Exercise. Consider the probabilistic model of Chapter 1, Example 12, where $\Omega = \{(x,y)|x^2 + y^2 \leqslant 1\}$, $P(A) = (1/\pi)S(A)$ and X and Y are defined as in Fig. 9. What is the probability density of the two-dimensional random vector $Z = (X, Y)$? What is the (marginal) density of X? Define $U = \sqrt{X^2 + Y^2}$ and $\Theta = \text{Arg}(X + iY)$ (see Fig. 9). What is the density of (U, Θ)?

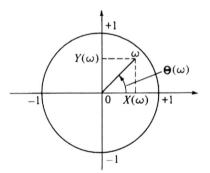

Figure 9. A random point in the unit disk.

Remark. A density probability f_X can always be modified in such a way that whenever $f_Y(y_1,\ldots,y_l) = 0$, then $f_X(y_1,\ldots,y_l,t_{l+1},\ldots,t_n) = 0$ for all $t_{l+1}, \ldots, t_n \in \mathbb{R}$. Although this result cannot be proved without recourse to measure theory, it is a technical subtlety that is not of great practical importance since all probability densities occurring in specific calculations will exhibit this property.

Expectation. Let $g : \mathbb{R}^n \to \mathbb{R}$ be a function such that $\int_{-\infty}^{+\infty} \ldots \int_{-\infty}^{+\infty} g(x_1,\ldots,x_n)$ $f_X(x_1,\ldots,x_n)\,dx_1 \ldots dx_n$ has a meaning. The expectation of the random variable $g(X_1, \ldots, X_n)$, denoted by $E[g(X_1, \ldots, X_n)]$, is then defined by

$$E[g(X_1, \ldots, X_n)] = \int_{-\infty}^{+\infty} \cdots \int_{-\infty}^{+\infty} g(x_1, \ldots, x_n) f_X(x_1, \ldots, x_n) \, dx_1 \ldots dx_n. \tag{44}$$

2.2. Covariance, Cross-Covariance, and Correlation

Mean Vector and Covariance Matrix. Let $X = (X_1, \ldots, X_n)$ be an n-dimensional random vector. Its *mean m* is the vector

$$m = \begin{pmatrix} m_1 \\ \vdots \\ m_n \end{pmatrix} = \begin{pmatrix} E[X_1] \\ \vdots \\ E[X_n] \end{pmatrix} \tag{45}$$

and its *covariance matrix* is the square matrix

$$\Gamma = \begin{pmatrix} \sigma_{11} & \cdots & \sigma_{1n} \\ \vdots & & \vdots \\ \sigma_{n1} & \cdots & \sigma_{nn} \end{pmatrix} \tag{46}$$

where

$$\sigma_{ij} = E[(X_i - m_i)(X_j - m_j)]. \tag{47}$$

In vector notation, denoting transposition by a "prime" symbol,

$$m = E[X], \quad \Gamma = E[(X - m)(X - m)']. \tag{48}$$

It is clear from Eq. (47) that the covariance matrix Γ is symmetric. We will now show that it is nonnegative (notation: $\Gamma \geqslant 0$) in the sense that, for all vectors $u = (u_1, \ldots, u_n) \in \mathbb{R}^n$,

$$u'\Gamma u \geqslant 0. \tag{49}$$

Indeed, $u'\Gamma u = \sum_{i=1}^{n} \sum_{j=1}^{n} u_i u_j \sigma_{ij} = E\{[\sum_{i=1}^{n} u_i(X_i - m_i)]^2\} \geqslant 0$.

The Cross-Covariance Matrix of Two Random Vectors. Let $X = (X_1, \ldots, X_n)$ and $Y = (Y_1, \ldots, Y_p)$ be two random vectors of mean m_X and m_Y, respectively. The cross-covariance matrix of X and Y is, by definition, the $n \times p$ matrix

$$\Sigma_{XY} = \{\sigma_{X_i Y_j}\}, \tag{50}$$

where

$$\sigma_{X_i Y_j} = E[(X_i - m_{X_i})(Y_j - m_{Y_j})]. \tag{51}$$

In abridged notation,

$$\Sigma_{XY} = E[(X - m_X)(Y - m_Y)']. \tag{52}$$

In particular $\Sigma_{XX} = \Gamma_X$, the covariance matrix of X. Obviously

$$\Sigma_{XY} = \Sigma'_{YX}. \tag{53}$$

We will now see how affine transformations on random vectors affect their mean vectors and cross-covariance matrix.

Let A be an $k \times n$ matrix, C be a $l \times p$ matrix, and b and d be two vectors of dimensions k and l, respectively. A straightforward computation shows that

$$m_{AX+b} = Am_X + b \tag{54}$$

and that

$$\Sigma_{AX+b, CY+d} = A\Sigma_{XY}C'. \tag{55}$$

In particular,

$$\Gamma_{AX+b} = A\Gamma_X A'. \tag{56}$$

Correlation. Two random vectors are said to be *uncorrelated* if $\Sigma_{XY} = 0$. When this is the case, the covariance matrix Γ_Z of Z where $Z' = (X', Y') \equiv (X_1, \ldots, X_n, Y_1, \ldots, Y_p)$ takes the block diagonal form

$$\Gamma_Z = \left(\begin{array}{c|c} \Sigma_{XX} & 0 \\ \hline 0 & \Sigma_{YY} \end{array} \right) = \left(\begin{array}{c|c} \Gamma_X & 0 \\ \hline 0 & \Gamma_Y \end{array} \right). \tag{57}$$

The Coefficient of Correlation of Two Random Variables. Let X and Y be two random variables with means m_X and m_Y and finite variances σ_X^2 and σ_Y^2. The covariance of X and Y is, by definition, the number

$$\sigma_{XY} = E[(X - m_X)(Y - m_Y)].$$

If $\sigma_X^2 > 0$ and $\sigma_Y^2 > 0$, one also defines the *coefficient of correlation* ρ_{XY} of X and Y:

$$\rho_{XY} = \frac{\sigma_{XY}}{\sigma_X \sigma_Y}. \tag{58}$$

When either $\sigma_X = 0$ or $\sigma_Y = 0$, one sets $\rho_{XY} = 0$. The correlation coefficient ρ_{XY} satisfies

$$\boxed{-1 \leqslant \rho_{XY} \leqslant +1}. \tag{59}$$

If $\rho_{XY} = 0$, X and Y are said to be *uncorrelated*. If $|\rho_{XY}| = 1$, there exist real numbers a and b, not both null, such that

$$a(X - m_X) = b(Y - m_Y), \quad \text{P-as.} \tag{60}$$

PROOF OF EQ. (59). Consider for all $\lambda \in \mathbb{R}$ the random variable $Z_\lambda = (X - m_X) + \lambda(Y - m_Y)$. We have $E[Z_\lambda^2] = \sigma_X^2 + 2\lambda\sigma_{XY} + \lambda^2\sigma_Y^2 \geqslant 0$. This being true for all $\lambda \in \mathbb{R}$, necessarily $\Delta = \sigma_{XY}^2 - \sigma_X^2\sigma_Y^2 \leqslant 0$, i.e., $|\rho_{XY}| \leqslant 1$. Suppose now that $|\rho_{XY}| = 1$. If $\sigma_X^2 = 0$, then $X - m_X = 0$, P-as, so that Eq. (60) holds with $a = 1$, $b = 0$. Therefore, suppose without loss of generality that $\sigma_X^2 > 0$ and $\sigma_Y^2 > 0$. From $|\rho_{XY}| = 1$, we obtain $\Delta = 0$. This implies the existence of a double root $\lambda_0 \in \mathbb{R}$ for $\sigma_X^2 + 2\lambda\sigma_{XY} + \lambda^2\sigma_Y^2$. Therefore, for such λ_0, $E[Z_{\lambda_0}^2] = 0$, which implies $Z_{\lambda_0} = 0$, P-as, that is, $(X - m_X) + \lambda_0(Y - m_Y) = 0$, P-as. $\square$

E12 Exercise. Let Θ be a random variable uniformly distributed on $[0, 2\pi]$. Show that $X = \sin\Theta$ and $Y = \cos\Theta$ are uncorrelated.

E13 Exercise. Consider the probabilistic model of E11. Compute the coefficient of correlation of X and Y.

E14 Exercise. Show that for all random variables X and Y with finite variance, and for all real numbers a, b, c, d

$$\rho_{aX+b, cY+d} = \rho_{X,Y}\operatorname{sgn}(ac) \tag{61}$$

where $\operatorname{sgn}(ac) = 1$ if $ac > 0$, -1 if $ac < 0$, and 0 if $ac = 0$.

Standardization of a Nondegenerate Random Vector. Let $X = (X_1, \ldots, X_n)$ be a random vector with mean m and covariance matrix $\Sigma_{XX} = \Gamma$. We say that X is nondegenerate iff

$$\Gamma > 0, \tag{62}$$

that is, iff $u \in \mathbb{R}^n$ and $u'\Gamma u = 0$ imply $u = 0$. It is proved in Linear Algebra courses that there exists an invertible $n \times n$ matrix B such that

$$\Gamma = BB'. \tag{63}$$

Taking $A = B^{-1}$ and $b = B^{-1}m_X$ in Eq. (56) we see that the vector $Y = B^{-1}(X - m_X)$ is such that $\Gamma_Y = B^{-1}BB'(B^{-1})' = I_n$ where I_n is the $(n \times n)$ identity matrix. Also the mean of Y is $m_Y = A^{-1}(m_X - m_X) = 0$. In summary,

$$X = BY + m_X \tag{64}$$

where Y is an n-dimensional vector with mean 0 and covariance matrix I_n. Such a vector Y will be called a *standard random vector*.

The standardization operation in the case of a random variable X *with mean* m_X and finite variance $\sigma_X^2 > 0$ reduces to $X = \sigma_X Y + m_X$ where

$$Y = \frac{X - m_X}{\sigma_X}. \tag{65}$$

Standardization in the Degenerate Case. In the general case, where X might be degenerate, we know from Linear Algebra that there exists a unitary $(n \times n)$ matrix C, i.e., a matrix describing an orthonormal change of basis in $\mathbb{R}^n$, or equivalently, an invertible matrix such that

$$C^{-1} = C', \tag{66}$$

such that

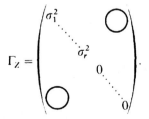

$$\Gamma_X = C \begin{pmatrix} \sigma_1^2 & & & & \\ & \ddots & & \bigcirc & \\ & & \sigma_r^2 & & \\ & & & 0 & \\ & \bigcirc & & & \ddots \\ & & & & 0 \end{pmatrix} C', \tag{67}$$

where r is the rank of Γ_X and $\sigma_i > 0 \, (1 \leqslant i \leqslant r)$. This decomposition is possible because Γ_X is symmetric and nonnegative.

Remark. From Algebra, it is known that σ_i^2 is an eigenvalue of Γ_X for all i, $1 \leqslant i \leqslant r$, and that if $r < n$, 0 is also an eigenvalue. Moreover, in Eq. (67) one can take $\sigma_1^2 \geqslant \sigma_2^2 \geqslant \cdots \geqslant 0$ and for the matrix C' a matrix of eigenvectors of Γ_X with the following arrangement: if $\sigma_{i-1}^2 > \sigma_i^2 = \sigma_{i+1}^2 = \sigma_{k-1}^2 > \sigma_k^2$, the columns of C' numbered from i to $k-1$ consist of vectors forming an orthonormal basis of the null space of $\Gamma_X - \sigma_i^2 I$. Note that B in Eq. (63) can be taken to be $C \, \text{diag}\{\sigma_1, \ldots, \sigma_n\}$ (if $\Gamma_X > 0$, $r = n$).

By Eqs. (66) and (67), if we define

$$Z = C^{-1}(X - m_X),$$

we see using Eq. (56) that $Z = (Z_1, \ldots, Z_n)$ is an n-vector with mean 0 and covariance matrix

$$\Gamma_Z = \begin{pmatrix} \sigma_1^2 & & & & \\ & \ddots & & \bigcirc & \\ & & \sigma_r^2 & & \\ & & & 0 & \\ & \bigcirc & & & \ddots \\ & & & & 0 \end{pmatrix}.$$

In particular, $Z_{r+i} = 0$, P-as, for all i, $1 \leqslant i \leqslant n - r$, or $Z = (Z_1, \ldots, Z_r, 0,$

..., 0), or equivalently,

$$\begin{pmatrix} Z_1 \\ \vdots \\ Z_n \end{pmatrix} = \begin{pmatrix} 1 & & & \bigcirc \\ & 1 & & \\ & & \ddots & \\ \bigcirc & & & 1 \\ \hline 0\,0 & \cdots & 0 \end{pmatrix} \begin{pmatrix} Z_1 \\ \vdots \\ Z_r \end{pmatrix}.$$

Now $Y = (Y_1, \ldots, Y_r)$ defined by

$$Y = \begin{pmatrix} \dfrac{1}{\sigma_1} & & \bigcirc \\ & \ddots & \\ \bigcirc & & \dfrac{1}{\sigma_r} \end{pmatrix} \begin{pmatrix} Z_1 \\ \vdots \\ Z_r \end{pmatrix}$$

is a standard r-vector. Therefore,

$$X = C \begin{pmatrix} 1 & & & \bigcirc \\ & 1 & & \\ & & \ddots & \\ \bigcirc & & & 1 \\ \hline 0\,0 & \cdots & 0 \end{pmatrix} \begin{pmatrix} \sigma_1 & & \bigcirc \\ & \ddots & \\ \bigcirc & & \sigma_r \end{pmatrix} Y + m_X$$

where Y is a standard vector of dimension r, the rank of Γ_X, and C is a unitary $n \times n$ matrix. In summary

$$X = BY + m_X \tag{68}$$

where Y is a standard vector of dimension r, and B is an $n \times r$ matrix of rank r.

Remark. If the rank of Γ_X is strictly less than n, the samples $X(\omega)$ of the random vector X lie almost surely in an affine hyperplane of dimension r of the affine space $\mathbb{R}^n$. We shall call r the *true dimension* of the random vector X (see Fig. 10).

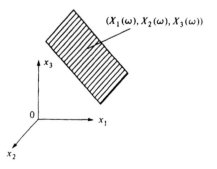

Figure 10. A 3-vector with true dimension 2.

E15 Exercise. Let $X = (X_1, X_2, X_3)$ be a random 3-vector with mean 0 and covariance matrix

$$\Gamma = \begin{pmatrix} 1 & 1 & 2 \\ 1 & 2 & 3 \\ 2 & 3 & 5 \end{pmatrix}.$$

Find a representation of X as $X = BY$ where Y is a standard vector of appropriate dimension.

2.3. Characteristic Function of a Random Vector

The characteristic function of a random vector $X = (X_1, \ldots, X_n)$ is the function $\phi_X : \mathbb{R}_n \to \mathbb{C}$ defined by

$$\phi_X(u_1, \ldots, u_n) = E[e^{i(u_1 X_1 + \cdots + u_n X_n)}], \tag{69}$$

that is, if X admits a density f_X,

$$\phi_X(u_1, \ldots, u_n) = \int_{-\infty}^{+\infty} \cdots \int_{-\infty}^{+\infty} e^{i(u_1 x_1 + \cdots + u_n x_n)} f_X(x_1, \ldots, x_n)\, dx_1 \ldots dx_n. \tag{70}$$

Note that for any i $(1 \leqslant i \leqslant n)$, the characteristic function of the random variable X_i is obtained as follows:

$$\phi_{X_i}(u) = \phi_X(0, 0, \ldots, 0, u, \ldots, 0), \tag{71}$$

where the u is in the ith position. More generally,

E16 Exercise. Show that

$$\phi_{\lambda_j X_j + \lambda_k X_k}(u) = \phi_X(0, \ldots, 0, \lambda_j u, 0, \ldots, 0, \lambda_k u, 0, \ldots, 0). \tag{72}$$

As we already saw in the univariate case, the characteristic function determines the distribution function.

When formally differentiating under the integration sign in Eq. (70) one obtains

$$\frac{\partial^k \phi_X}{\partial u_1^{k_1} \ldots \partial u_n^{k_n}}(u_1, \ldots, u_k)$$

$$= \int_{-\infty}^{+\infty} \cdots \int_{-\infty}^{+\infty} (i)^k x_1^{k_1} \ldots x_n^{k_n} e^{i(u_1 x_1 + \cdots + u_n x_n)} f_X(x_1, \ldots, x_n)\, dx_1 \ldots dx_n$$

where $k = k_1 + \cdots + k_n$. Therefore

$$\frac{\partial^k \phi_X}{\partial u_1^{k_1} \ldots \partial u_n^{k_n}}(0,\ldots,0) = (i)^k E[X_1^{k_1} \ldots X_n^{k_n}] \qquad (73)$$

This formula will be justified in Section 4 and is valid whenever $E[|X_1|^{k_1} \ldots |X_n|^{k_n}] < \infty$.

E17 Exercise. Compute $E[X^n]$ when $X \sim \mathscr{E}(\lambda)$.

Characteristic functions are very useful in asserting or disproving independence as we will see in the next section.

3. Independence

3.1. Independent Random Variables

Product of Densities. The real random variables $X_1,\ldots,X_n$ are said to be independent iff for all $x_1,\ldots,x_n \in \mathbb{R}$

$$P(X_1 \leqslant x_1,\ldots,X_n \leqslant x_n) = P(X_1 \leqslant x_1)\ldots P(X_n \leqslant x_n). \qquad (74)$$

Suppose, moreover, that each X_i admits a density f_i. Then, letting $X = (X_1,\ldots,X_n)$, we have

$$F_X(x_1,\ldots,x_n) = \left(\int_{-\infty}^{x_1} f_1(y_1)\,dy_1\right)\ldots\left(\int_{-\infty}^{x_n} f_n(y_n)\,dy_n\right)$$

$$= \int_{-\infty}^{x_1} \ldots \int_{-\infty}^{x_n} f_1(y_1)\ldots f_n(y_n)\,dy_1\ldots dy_n$$

where we have used Fubini's theorem. In other words, X admits a density f_X which is the product of the (marginal) densities

$$f_X(x_1,\ldots,x_n) = \prod_{i=1}^{n} f_i(x_i) \qquad (75)$$

The converse is also true: if the density f_X of $X = (X_1,\ldots,X_n)$ factors as in Eq. (75) where $f_1,\ldots,f_n$ are univariate probability densities, then $X_1,\ldots,X_n$ are independent random variables admitting the probability densities $f_1,\ldots,f_n$, respectively. Indeed, by Fubini,

$$P(X_1 \leqslant x_1) = P(X_1 \leqslant x_1, X_2 < \infty, \ldots, X_n < \infty)$$

$$= \int_{-\infty}^{x_1} \int_{-\infty}^{+\infty} \ldots \int_{-\infty}^{+\infty} f_1(y_1)f_2(y_2) \ldots f_n(y_n)\, dy_1\, dy_2 \ldots dy_n$$

$$= \left(\int_{-\infty}^{x_1} f_1(y_1)\, dy_1\right)\left(\int_{-\infty}^{+\infty} f_2(y_2)\, dy_2\right) \ldots \left(\int_{-\infty}^{+\infty} f_n(y_n)\, dy_n\right)$$

$$= \int_{-\infty}^{x_1} f_1(y_1)\, dy_1,$$

and therefore f_1 is the density of X_1. Similarly, for all i ($1 \leqslant i \leqslant n$), f_i is the density of X_i. Now, again by Fubini,

$$P(X_1 \leqslant x_1, \ldots, X_n \leqslant x_n) = \int_{-\infty}^{x_1} \ldots \int_{-\infty}^{x_n} f_1(y_1) \ldots f_n(y_n)\, dy_1 \ldots dy_n$$

$$= \left(\int_{-\infty}^{x_1} f_1(y_1)\, dy_1\right) \ldots \left(\int_{-\infty}^{x_n} f_n(y_n)\, dy_n\right)$$

$$= P(X_1 \leqslant x_1) \ldots P(X_n \leqslant x_n),$$

which proves the independence of the X_i's.

Product of Expectations. Let $X_1, \ldots, X_n$ be independent real random variables admitting the probability densities $f_1, \ldots, f_n$, respectively. Then, for any functions $g_1, \ldots, g_n$ from $\mathbb{R}$ into $\mathbb{R}$

$$\boxed{E\left[\prod_{i=1}^{n} g_i(X_i)\right] = \prod_{i=1}^{n} E[g_i(X_i)]} \tag{76}$$

as long as the expectations involved have a meaning. This is a general result that is valid even when no probability density is available for the X_i's (see Section 4).

PROOF. The proof in the density case is a straightforward application of Fubini's theorem:

$$\int_{-\infty}^{+\infty} \ldots \int_{-\infty}^{+\infty} g_1(x_1) \ldots g_n(x_n)f_1(x_1) \ldots f_n(x_n)\, dx_1 \ldots dx_n$$

$$= \prod_{i=1}^{n} \int_{-\infty}^{+\infty} g_i(x_i)f_i(x_i)\, dx_i. \qquad \square$$

E18 Exercise (E11 continued). Show that U and Θ are independent and that X and Y are not independent.

Product of Characteristic Functions. The product formula (76) is also true when the g_i's are complex functions, as can be readily checked. In particular, if $X_1, \ldots, X_n$ are independent

$$\phi_{X_1,\ldots,X_n}(u_1,\ldots,u_n) = \prod_{i=1}^{n} \phi_{X_i}(u_i). \tag{77}$$

The converse is also true, but we cannot prove it at this stage. If Eq. (77) holds for all $u = (u_1, \ldots, u_n) \in \mathbb{R}^n$, then $X_1, \ldots, X_n$ are independent.

E19 Exercise. Let $X_1, \ldots, X_n$ be n independent Cauchy random variables. Recall (see Exercise E9) that the characteristic function of a Cauchy random variable is $\phi(u) = e^{-|u|}$. Show that $Z_n = (X_1 + \cdots + X_n)/n$ is also a Cauchy random variable.

E20 Exercise. Let X_1 and X_2 be two independent random variables such that $X_1 \sim \gamma(\alpha_1, \beta)$ and $X_2 \sim \gamma(\alpha_2, \beta)$ where $\alpha_1 > 0$, $\alpha_2 > 0$, $\beta > 0$. Show that $X_1 + X_2 \sim \gamma(\alpha_1 + \alpha_2, \beta)$. (Hint: Use the result of Exercise E7.)

Variance of a Sum of Independent Random Variables. Let $X_1, \ldots, X_n$ be random variables with means $m_1, \ldots, m_n$ and variances $\sigma_1^2, \ldots, \sigma_n^2$, respectively. Let m and σ^2 be the mean and variance of $Z = X_1 + \cdots + X_n$. We already know that $m = m_1 + \cdots + m_n$ (and this is true whether the X_i's are independent or not). If in addition $X_1, \ldots, X_n$ are independent, then

$$\boxed{\sigma^2 = \sigma_1^2 + \cdots + \sigma_n^2}. \tag{78}$$

E21 Exercise. Prove Eq. (78).

The Convolution Formula. Let X and Y be two independent real random variables with the probability densities f_X and f_Y, respectively. Then, $Z = X + Y$ admits a density f_Z given by

$$\boxed{f_Z(z) = \int_{-\infty}^{+\infty} f_Y(z - x) f_X(x)\, dx = \int_{-\infty}^{+\infty} f_X(z - y) f_Y(y)\, dy}. \tag{79}$$

In other words, f_Z is the convolution of f_X and f_Y. This is denoted by $f_Z = f_X * f_Y$.

Proof. $P(Z \leqslant z) = P(X + Y \leqslant z) = P(A) = E[1_A]$ where $A = \{\omega \,|\, X(\omega) + Y(\omega) \leqslant z\}$. But $1_A = g(X, Y)$ where $g(X(\omega), Y(\omega)) = 1_C(X(\omega), Y(\omega))$ and $C = \{(x, y) \,|\, x + y \leqslant z\}$. Therefore, $E[1_A] = E[g(X, Y)] = \int_{-\infty}^{+\infty} \int_{-\infty}^{+\infty} g(x, y) \cdot f_X(x) f_Y(y)\, dx\, dy = \int_{-\infty}^{+\infty} \int_{-\infty}^{+\infty} 1_C(x, y) f_X(x) f_Y(y)\, dx\, dy = \iint_C f_X(x) f_Y(y)\, dx\, dy = \int_{-\infty}^{+\infty} (\int_{-\infty}^{z-x} f_Y(y)\, dy) f_X(x)\, dx = \int_{-\infty}^{+\infty} (\int_{-\infty}^{z} f_Y(y - x)\, dy) f_X(x)\, dx = \int_{-\infty}^{z} [\int_{-\infty}^{+\infty} f_Y(y - x) f_X(x)\, dx]\, dy$. Therefore, $f_Z(z)$ defined by Eq. (79) verifies $P(Z \leqslant z) = \int_{-\infty}^{z} f_Z(y)\, dy$. $\qquad\square$

E22 Exercise. Let X and Y be two independent random variables uniformly distributed over $[a, b]$ and $[c, d]$, respectively. Show that $Z = X + Y$ has the following probability density in the case where $b - a \leqslant d - c$ (Fig. 11).

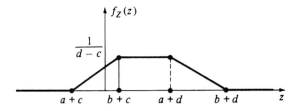

Figure 11. The p.d. of the sum of two random variables uniformly distributed on intervals.

3.2. Independent Random Vectors

For notational ease, we shall state the results for *two* vectors $X = (X_1, \ldots, X_n)$ and $Y = (Y_1, \ldots, Y_p)$. In complete analogy with the univariate case, we say that X and Y are independent if and only if for all $x \in \mathbb{R}^n$, $y \in \mathbb{R}^p$

$$P(X \leqslant x, Y \leqslant y) = P(X \leqslant x)P(Y \leqslant y)$$ (80)

[Recall that when $x, a \in \mathbb{R}^k$, $x \leqslant a$ means $x_1 \leqslant a_1, \ldots, x_k \leqslant a_k$]. If, moreover, X and Y admit the probability densities f_X and f_Y, then $Z = (X, Y) = (X_1, \ldots, X_n, Y_1, \ldots, Y_p)$ admits a probability density f_Z given by

$$f_Z(x, y) = f_X(x)f_Y(y)$$ (81)

Conversely, if X and Y are two vectors such that $Z = (X, Y)$ admits a density f_Z that can be factored as

$$f_Z(x, y) = f_1(x)f_2(y),$$

where f_1 and f_2 are probability densities, then X and Y are independent and admit the probability densities $f_X = f_1$ and $f_Y = f_2$, respectively. The proof is exactly the same as in the univariate case. Similarly, a necessary and sufficient condition for vectors X and Y to be independent is that for all $u \in \mathbb{R}^n$, $v \in \mathbb{R}^p$,

$$E[e^{i(u'X + v'Y)}] = E[e^{iu'X}]E[e^{iv'Y}],$$

or, in obvious notations,

$$\phi_{X,Y}(u, v) = \phi_X(u)\phi_Y(v)$$ (82)

Also, if g and h are functions from $\mathbb{R}^n$ and $\mathbb{R}^p$, respectively, into $\mathbb{R}$, and if X and Y are independent, then

$$E[g(X)h(Y)] = E[g(X)]E[h(Y)] \tag{83}$$

provided the quantities featured in the above equality have a meaning. From this, it is not hard to see that when the covariance and cross-covariance matrices exist, and if X and Y are independent,

$$\Gamma_{X+Y} = \Gamma_X + \Gamma_Y, \quad \Sigma_{XY} = 0 \tag{84}$$

In particular, the covariance matrix of $Z = (X, Y) = (X_1, \ldots, X_n, Y_1, \ldots, Y_p)$ takes the block diagonal form

$$\Gamma_Z = \left(\begin{array}{c|c} \Gamma_X & 0 \\ \hline 0 & \Gamma_Y \end{array} \right). \tag{85}$$

It follows from the product theorem that if $X_1, \ldots, X_n$ are independent random vectors of dimensions $k_1, \ldots, k_n$, respectively, then for all $C_1 \in \mathscr{B}^{k_1}, \ldots, C_n \in \mathscr{B}^{k_n}$,

$$P(X_1 \in C_1, \ldots, X_n \in C_n) = P(X_1 \in C_1) \ldots P(X_n \in C_n), \tag{86}$$

and therefore that this statement is equivalent to independence.

E23 Exercise. Prove Eq. (86).

Now for $1 \leqslant i \leqslant n$, let g_i be a function from $\mathbb{R}^{k_i}$ into $\mathbb{R}^{m_i}$ and let

$$Y_i = g_i(X_i) \quad (1 \leqslant i \leqslant n). \tag{87}$$

The random vectors $Y_1, \ldots, Y_n$ are then independent. To prove this, it suffices to show that, for all $D_i \in \mathscr{B}^{m_i}$ $(1 \leqslant i \leqslant n)$,

$$P(Y_1 \in D_1, \ldots, Y_n \in D_n) = P(Y_1 \in D_1) \ldots P(Y_n \in D_n).$$

Since $\{g_i(X_i) \in D_i\} = \{X_i \in g_i^{-1}(D_i)\}$ $(1 \leqslant i \leqslant n)$, the last statement is equivalent to Eq. (86) with $C_i = g_i^{-1}(D_i)$ $(1 \leqslant i \leqslant n)$.

4. Random Variables That Are Not Discrete and Do Not Have a Probability Density

This textbook is an *introduction* to the basic concepts of Probability Theory. As such it avoids the most technical aspects of the theory since the mathematical notions required for a complete and rigorous treatment are generally taught at the graduate level. This is a minor inconvenience for many applications of Probability Theory in the applied sciences. However, recourse to a mathematically more sophisticated approach cannot always be avoided. One objective of this section is to motivate the reader to further study, especially of Integration Theory, which is the foundation of Probability Theory. Another objective is to fill a few gaps that we have been forced to leave open and that, in some cases, may have disturbed the reader.

Virtually no proof of the results stated in this section will be provided for that is the object of a full course in Mathematics of intrinsic interest. We have chosen two themes, the general definition of expectation and Lebesgue's theorems, because they allow us to prove a few results that we previously had to accept on faith.

4.1. The Abstract Definition of Expectation

First it must be said that not all random variables are either discrete or have a probability density. It is not difficult to produce a counterexample: indeed, let U and Y be two independent random variables with $Y \sim \mathcal{N}(0,1)$ and $P(U = 0) = P(U = 1) = \frac{1}{2}$. The random variable $X = UY$ does not admit a probability density since $P(X = 0) = P(U = 0) = \frac{1}{2}$, and it is not discrete since $P(X \in [a,b] - \{0\}) = \frac{1}{2}(1/\sqrt{2\pi})\int_a^b \exp(-x^2/2)\,dx$ for all a, $b \in \mathbb{R}$ such that $a \leq b$.

The above counterexample cannot be called pathological, and in fact, the most general random variables have distribution functions that do not allow a definition of expectation as simple as in the case of random variables which are either discrete or with a probability density. One must therefore resort to a more general (and more abstract) definition of expectation if the theory is not to be unacceptably incomplete.

Such a definition is available. More precisely, let X be a nonnegative random variable. Its expectation, denoted $E[X]$, is defined by

$$E[X] = \lim_{n \uparrow \infty} \uparrow \sum_{k=0}^{n2^n - 1} \frac{k}{2^n} P\left(\frac{k}{2^n} \leq X < \frac{k+1}{2^n}\right) + nP(X \geq n) . \tag{88}$$

Of course, this definition coincides with the special definitions in the discrete case and in the case where a probability density exists. In the probability density case, for instance, Eq. (88) reads

$$E[X] = \lim_{n \uparrow \infty} \sum_{k=0}^{n2^n - 1} \frac{k}{2^n} \int_{k/2^n}^{(k+1)/2^n} f(y)\,dy + n \int_n^\infty f(y)\,dy$$

$$= \lim_{n \uparrow \infty} \int_0^\infty \left\{ \sum_{k=0}^{n2^n - 1} \frac{k}{2^n} 1_{[k/2^n, (k+1)/2^n)}(y) + n1_{[n, \infty)}(y) \right\} f(y)\,dy.$$

If f is continuous, or piecewise continuous, it can be directly checked that the latter expression is just $\int_0^\infty yf(y)\,dy$. (It is also a consequence of Lebesgue's monotone convergence theorem, to be stated in Section 4.2.)

For a random variable X that is not nonnegative, the procedure already used to define $E[X]$ in the discrete case and in the probability density case is still applicable, i.e., $E[X] = E[X^+] - E[X^-]$ if not both $E[X^+]$ and $E[X^-]$ are infinite. If $E[X^+]$ *and* $E[X^-]$ are infinite, the expectation is not defined. If $E[|X|] < \infty$, X is said to be integrable, and then $E[X]$ is a finite number.

The basic properties of the expectation so defined are linearity and monotonicity: if X_1 and X_2 are random variables with expectations, then for all λ_1, $\lambda_2 \in \mathbb{R}$,

$$E[\lambda_1 X_1 + \lambda_2 X_2] = \lambda_1 E[X_1] + \lambda_2 E[X_2] \tag{89}$$

whenever the right-hand side has a meaning (i.e., is not a $\infty - \infty$ form). Also, if $X_1 \leqslant X_2$, P-as

$$E[X_1] \leqslant E[X_2]. \tag{90}$$

Also from definition (88) applied to $X = 1_A$,

$$P(A) = E[1_A]. \tag{91}$$

The definitions of the mean, the variance, and the characteristic function of a random variable are the same as in the discrete case and the probability density case when written in terms of the $E[\cdots]$ symbolism:

$$\left\{ \begin{array}{ll} m_X = E[X] & \text{(92a)} \\[2mm] \sigma_X^2 = E[(X - m_X)^2] = E[X^2] - m_X^2 & \text{(92b)} \\[2mm] \phi_X(u) = E[e^{iuX}] & \text{(92c)} \end{array} \right.$$

where in Eq. (92c), $E[e^{iuX}] = E[\cos uX] + iE[\sin uX]$. In Section 1.3, a proof of the Chebyshev's inequality

$$P(|X - m_X| \geqslant \varepsilon) \leqslant \frac{\sigma_X^2}{\varepsilon^2} \tag{93}$$

was given in terms of the $E[\cdots]$ symbolism, only using properties (89), (90) and (91). Therefore, Eq. (93) is also true of any random variable with mean m_X and variance σ_X^2.

As a direct consequence of Eq. (88) we see that if $E|X| < \infty$

$$E[X] = \int_{-\infty}^{+\infty} x\,dF(x) \tag{94a}$$

where $F(x) = P(X \leqslant x)$ is the distribution function of X and where the integral in the right-hand side is the Riemann–Stieltjes integral:

$$\int_{-\infty}^{+\infty} x\,dF(x) = \lim_{n \uparrow \infty} \sum_{\substack{k \in \mathbb{Z} \\ |k| \leqslant n2^n - 1}} \frac{k}{2^n}\left(F\left(\frac{k+1}{2^n}\right) - F\left(\frac{k}{2^n}\right)\right). \tag{$*$}$$

More generally, we have the formula:

$$E[g(X)] = \int_{-\infty}^{+\infty} g(x)\,dF(x)$$ (94b)

for any function g such that $E[|g(X)|] < \infty$.

At this point, we shall quote without proof a formula that is true whenever X is a *nonnegative* random variable:

$$E[X] = \int_0^{\infty} [1 - F(x)]\,dx .$$ (95)

A heuristic proof can be based on Fig. 12.

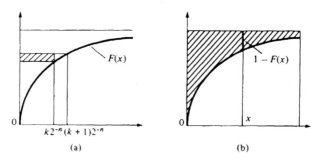

(a)   (b)

Figure 12. Proof of formula (95).

The shaded area in Fig. 12(a) is just $k2^{-n}(F((k + 1)2^{-n}) - F(k2^{-n}))$, and therefore, from Eqs. (∗) and (94a), we see that $E[X]$ is the shaded area in Fig. 12(b), which can also be expressed as $\int_0^{+\infty} (1 - F(x))\,dx$.

It was shown in Chapter 1, Section 3.2, that the c.d.f. F of a random variable is nondecreasing, right-continuous, with a left-hand limit at each point $x \in \mathbb{R}$, denoted $F(x-)$. Moreover $F(-\infty) = 0$ and $F(+\infty) = 1$. A point $x \in \mathbb{R}$ is called a *discontinuity point* of F if $\Delta F(x) > 0$ where $\Delta F(x) = F(x) - F(x-)$.

Exercise. The discontinuity points of F form a set which is either finite or denumerable.

Let $(x_n, n \geqslant 1)$ be an enumeration of the discontinuity points of F. It must be noted that the x_n's may have a (denumerable) infinity of accumulation points. The *continuous part* of F, denoted F_c, is defined by

$$F_c(x) = F(x) - \sum_{x_n \leqslant x} \Delta F(x_n).$$

The function F_c is continuous, but in general it may not be written as an integral

$$F_c(x) = \int_{-\infty}^{x} f_c(u)\,du.$$

However if this is the case

$$E[g(X)] = \int_{-\infty}^{+\infty} g(x)f_c(x)\,dx + \sum_{n \geq 1} g(x_n)\Delta F(x_n).$$

In Engineering texts, especially in Signal Theory, it is then said that X admits a *mixed density*

$$f(x) = f_c(x) + \sum_{n \geq 1} p_n\delta(x - x_n)$$

where

$$p_n = \Delta F(x_n) = P(X = x_n)$$

and δ is the so-called Dirac "function" formally defined by its action on any function $g : \mathbb{R} \to \mathbb{R}$:

$$\int_{-\infty}^{+\infty} g(x)\delta(x - x_0)\,dx = g(x_0).$$

This notation is consistent with the following definition of expectation using the mixed density f:

$$Eg(x) = \int_{-\infty}^{+\infty} g(x)f(x)\,dx.$$

Indeed

$$\int_{-\infty}^{+\infty} g(x)f(x)\,dx = \int_{-\infty}^{+\infty} g(x)\left[f_c(x) + \sum_{n \geq 1} p_n\delta(x - x_n)\right]dx$$

$$= \int_{-\infty}^{+\infty} g(x)f_c(x)\,dx + \sum_{n \geq 1} g(x_n)\Delta F(x_n).$$

The mixed densities are graphically represented as a mixture of "Dirac functions" and of ordinary functions (Fig. 13).

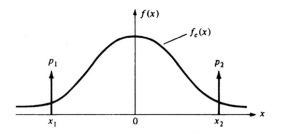

Figure 13. A mixed density.

We will now mention a difficulty that we have tried to conceal in the definition of expectation in the case of a random vector $X = (X_1, \ldots, X_n)$ with a probability density f_X. For instance, we have defined for $Y = g(X_1, \ldots, X_n)$ the expectation $E[Y]$ by

$$E[Y] = \int_{-\infty}^{+\infty} \cdots \int_{-\infty}^{+\infty} g(x_1, \ldots, x_n) f_X(x_1, \ldots, x_n)\, dx_1 \ldots dx_n.$$

But now, suppose that Y admits a probability density f_Y. Expectation $E[Y]$ could be computed by means of

$$E[Y] = \int_{-\infty}^{+\infty} y f_Y(y)\, dy.$$

The unavoidable question is: are the last two definitions consistent? Without the help of abstract Integration Theory, such a question cannot be answered in general. Fortunately, the answer is yes. The theory tells us that all the special definitions of expectation given in the present textbook are consistent with Eq. (88).

4.2. Lebesgue's Theorems and Applications

The most important technical results relative to expectation are the celebrated Lebesgue theorems, which give fairly general conditions under which the limit and expectation symbols can be interchanged, i.e.,

$$E\left[\lim_{n \uparrow \infty} X_n\right] = \lim_{n \uparrow \infty} E[X_n]. \tag{96}$$

Monotone Convergence Theorem (MC). *Let* $(X_n, n \geqslant 1)$ *be a sequence of random variables such that for all* $n \geqslant 1$,

$$0 \leqslant X_n \leqslant X_{n+1}, \quad \text{P-as.} \tag{97}$$

Then Eq. (96) *holds.*

Dominated Convergence Theorem (DC). *Let* $(X_n, n \geqslant 1)$ *be a sequence of random variables such that for all* ω *outside a set* $\mathcal{N}$ *of null probability, there exists* $\lim_{n \uparrow \infty} X_n(\omega)$* *and such that for all* $n \geqslant 1$

$$|X_n| \leqslant Y, \quad \text{P-as} \tag{98}$$

where Y *is some integrable random variable. Then Eq.* (96) *holds.*

A simple counterexample will show that Eq. (96) is not always true when $\lim_{n \uparrow \infty} X_n$ exists. Indeed, take the following probabilistic model: $\Omega = [0, 1]$, and P is the Lebesgue measure on $[0, 1]$. Thus, ω is a real number in $[0, 1]$,

---

* One then says that $(X_n, n \geqslant 1)$ converges almost surely. We shall spend more time on almost-sure convergence in Chapter 5.

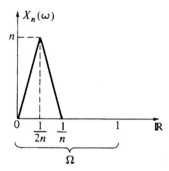

Figure 14. A counterexample for Lebesgue's theorems.

and a random variable can be assimilated to a real function defined on $[0,1]$. Take for X_n the function depicted in Fig. 14. Clearly, $\lim_{n\uparrow\infty} X_n(\omega) = 0$ and $E[X_n] = \int_0^1 X_n(x)\,dx = 1$, so that $E[\lim_{n\uparrow\infty} X_n] = 0 \neq \lim_{n\uparrow\infty} E[X_n] = 1$.

We shall now harvest a few applications of Lebesgue's theorems. The monotone convergence theorem yields the following result concerning a sequence of nonnegative random variables $(S_n, n \geq 1)$:

$$E\left[\sum_{n=1}^{\infty} S_n\right] = \sum_{n=1}^{\infty} E[S_n]. \tag{99}$$

It suffices to apply Lebesgue's MC theorem, with $X_n = \sum_{k=1}^{n} S_k$. This result has repeatedly been used without proof in the previous chapters.

Differentiation Under the Expectation Sign (DUE)

Let $(X_t, t \in [a,b])$ be a family of integrable real random variables such that for each ω, $t \to X_t(\omega)$ admits a derivative in (a,b) denoted $\dot{X}_t(\omega)$. Suppose, moreover, that there exists an integrable random variable Y such that $|\dot{X}_t| \leq Y$, P-as, for all $t \in (a,b)$. Then

$$\boxed{\frac{d}{dt} E[X_t] = E[\dot{X}_t]} \qquad [t \in (a,b)]. \tag{100}$$

PROOF. We have by definition of derivative $\dot{X}_t = \lim_{n\uparrow\infty} n(X_{t+(1/n)} - X_t)$. Since $n(X_{t+(1/n)} - X_t) = \dot{X}_{t+\theta(1/n)}$ for some $\theta \in (0,1)$, and since $|\dot{X}_t| \leq Y$, P-as, for all $t \in (a,b)$, DC is applicable. Therefore,

$$\lim_{n\uparrow\infty} E[n(X_{t+(1/n)} - X_t)] = E[\dot{X}_t].$$

But, using linearity and the definition of $(d/dt)E[X_t]$,

$$\lim_{n\uparrow\infty} E[n(X_{t+(1/n)} - X_t)] = \lim_{n\uparrow\infty} n(E[X_{t+(1/n)}] - E[X_t]) = \frac{d}{dt} E[X_t]. \quad \square$$

Application. Let X be a random variable with the characteristic function ϕ_X. Suppose that $E[|X|] < \infty$. Then

$$\frac{d\phi_X}{du}(u) = iE[Xe^{iuX}] \tag{101a}$$

and in particular

$$\frac{d\phi_X}{du}(0) = iE[X]. \tag{101b}$$

The proof of Eq. (101) is a simple application of DUE after noticing that $(d/du)e^{iuX} = iXe^{iuX}$. [In the notations of Eq. (100), $X_u = e^{iuX}$, $\dot{X}_u = iXe^{iuX}$, and therefore $|\dot{X}_u| \leqslant |X|$. Thus, DUE is applicable with $Y = |X|$.] Iteration of the differentiation process yields, under the condition $E[|X|^k] < \infty$,

$$\frac{d^k\phi_X}{du^k}(u) = (i)^kE[X^ke^{iuX}], \tag{102a}$$

and therefore

$$\boxed{\frac{d^k\phi_X}{du^k}(0) = i^kE[X^k]} \ . \tag{102b}$$

Remark. The condition $E[|X|^k] < \infty$ implies that $E[|X|^l] < \infty$ for all l $0 \leqslant l \leqslant k$. Indeed if $l \leqslant k$, and $a > 0$, $a^l \leqslant 1 + a^k$ and therefore $|X|^l \leqslant 1 + |X|^k$, so that $E[|X|^l] \leqslant 1 + E[|X|^k] < \infty$. This remark is necessary to perform the successive steps that lead to Eq. (102a).

The Product Formula

As another illustration of the monotone convergence theorem we shall now prove the product formula

$$\boxed{E[XY] = E[X]E[Y]} \tag{103}$$

which is true when X and Y are *independent* integrable random variables. Such a formula was proved in Section 3.2 using Fubini's theorem, for couples (X, Y) admitting a density. It was also proved in Chapter 2 in the discrete case using "elementary" arguments concerning series. However, a close look at the proof of the validity of such arguments would show that Lebesgue's theorems are already there.

The proof of Eq. (103) will be given here only in the case where X and Y are nonnegative. The rest of the proof is left to the reader.

First, notice that for any nonnegative random variable Z, if we define

Illustration 6. Buffon's Needle: A Problem in Random Geometry   117

$$Z_n = \sum_{k=0}^{n2^n-1} \frac{k}{2^n} 1_{\{(k/2^n) \leqslant Z < (k+1)/2^n\}} + n1_{\{Z \geqslant n\}} \qquad (104)$$

then $\lim_{n\uparrow\infty} \uparrow Z_n = Z$, P-as. Thus, by Lebesgue monotone convergence theorem, $\lim_{n\uparrow\infty} E[Z_n] = E[Z]$. With obvious notations, $X_n Y_n \uparrow XY$ and therefore $\lim_{n\uparrow\infty} E[X_n Y_n] = E[XY]$. Also, $\lim_{n\uparrow\infty} E[X_n]E[Y_n] = E[X]E[Y]$. It is then clear that Eq. (103) will be proved if

$$E[X_n Y_n] = E[X_n]E[Y_n]$$

is shown to be true for all $n \geqslant 1$. But then, this verification amounts to observing that for all i, j,

$$E[1_{\{(j/2^n) < X \leqslant (j+1)/2^n\}} 1_{\{(i/2^n) < Y \leqslant (i+1)/2^n\}}]$$
$$= E[1_{\{(j/2^n) < X \leqslant (j+1)/2^n\}}] E[1_{\{(i/2^n) < Y \leqslant (i+1)/2^n\}}].$$

That is, in view of Eq. (91),

$$P\left(X \in \left(\frac{j}{2^n}, \frac{j+1}{2^n}\right], Y \in \left(\frac{i}{2^n}, \frac{i+1}{2^n}\right]\right)$$
$$= P\left(X \in \left(\frac{j}{2^n}, \frac{j+1}{2^n}\right]\right) P\left(Y \in \left(\frac{i}{2^n}, \frac{i+1}{2^n}\right]\right).$$

The latter equality is an expression of the independence of X and Y, and the proof is completed.

Illustration 6. Buffon's Needle: A Problem in Random Geometry

The French naturalist Buffon gave his name to the famous problem of computing the probability that a needle of length l intersects a regular grid of parallel straight lines separated by a distance a, where $a > l$ (Fig. 15).

We will consider a more general problem, that of computing the average number of intersections of a "regular" plane curve $\overset{\frown}{AB}$. Here the length l of the curve and the grid's characteristic distance a are not restricted (Fig. 16).

The original problem's solution immediately follows from that of the more general problem, since if we call N the (random) number of intersections of the needle with the array, we have, when $a > l$, $N = 0$ or 1, and therefore

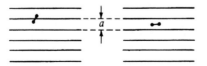

Figure 15. A needle on a grid.

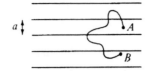

Figure 16. A piece of wire on a grid.

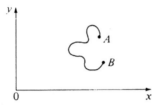

Figure 17. The piece of wire and its referential.

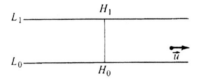

Figure 18. A referential of the grid.

$$E[N] = 0 \cdot P(N = 0) + 1 \cdot P(N = 1) = P(N = 1). \tag{105}$$

But $P(N = 1)$ is simply the probability that the needle intersects the grid.

Before solving the general Buffon's problem, one must describe in mathematical terms the operation of tossing a plane line "at random" on a plane surface. We propose the following model. The curve $\overset{\frown}{AB}$ must be thought of as a piece of wire stuck on a transparent sheet of plastic with an orthonormal referential $0xy$ drawn on it (Fig. 17). Two consecutive lines of the array are selected, e.g., L_0 and L_1, as well as two points H_0 and H_1 on L_0 and L_1, respectively, such that segment $H_0 H_1$ is perpendicular to the lines of the grid. Also let $\vec{u}$ be a fixed vector, e.g., parallel to the lines and from left to right, as in Fig. 18. Let M be a random point on $H_0 H_1$ such that, if we call X the random variable $H_0 M$,

$$P(X \in [c, d]) = (d - c)/a \tag{106}$$

for all $0 \leqslant c \leqslant d \leqslant a$. In other words, X is uniformly distributed on $[0, a]$.

We place the transparent plastic sheet on the rectangular array in such a manner that 0 coincides with the random point M, and we give to $\overrightarrow{0x}$ a random direction, i.e., the angle from direction $\vec{u}$ to direction $\overrightarrow{0x}$ is a random variable $\textcircled{H}$ such that

Illustration 6. Buffon's Needle: A Problem in Random Geometry   119

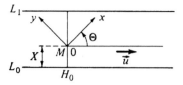

Figure 19. Random placement of the piece of wire.

$$P(\textcircled{H} \in [\theta_1, \theta_2]) = \frac{\theta_2 - \theta_1}{2\pi} \tag{107}$$

for all $0 \leqslant \theta_1 \leqslant \theta_2 \leqslant 2\pi$. The angle $\textcircled{H}$ is uniformly distributed between 0 and 2π (Fig. 19). We will assume that X and $\textcircled{H}$ are independent, i.e.,

$$P(X \in [c,d], \textcircled{H} \in [\theta_1, \theta_2]) = P(X \in [c,d])P(\textcircled{H} \in [\theta_1, \theta_2]). \tag{108}$$

This placement procedure of the transparent sheet, and therefore of the piece of wire $\overset{\frown}{AB}$, may not appear very random because of the particular choices of lines L_0, L_1 and of points H_0, H_1. However, the reader will agree that the distribution of the random number of intersections N of the line $\overset{\frown}{AB}$ with the array does not depend on such a choice.

The reader may not agree on the choice of Eqs. (106), (107), and (108) for the joint distribution of $\textcircled{H}$ and X. There is nothing we can say in favor of this choice except that it seems to be a reasonable model. Other models are mathematically acceptable. We are unable to determine which model represents accurately the actual throwing of the piece of wire if no description is available of the physical apparatus designed to throw the piece of wire. We will consider only those tossing machines that achieve Eqs. (106), (107), and (108), and we will be content with them.

We now proceed to the solution of Buffon's needle problem in three steps.

Step 1. The average number of intersections of a segment of length l with the array is a function of l, say $f(l)$, to be determined. Consider any such segment AB and let C be some point on AB. Call l_1 and l_2 the length of AC and CB, respectively. Clearly the number of intersections of AB with the grid is equal to the sum of the number of intersections of its constituents AC and CB, so that $f(l) = f(l_1) + f(l_2)$. We have just obtained for f the functional equation

$$f(l_1 + l_2) = f(l_1) + f(l_2). \tag{109}$$

Therefore,

$$f(l) = kl \tag{110}$$

where k is a constant which will be computed in Step 3.

Step 2. Now let $M_0, M_1 \dots M_n$ be a broken line approximating $\overset{\frown}{AB}$ (Fig. 20), and let $l_1, \dots, l_n$ be the lengths of $M_0 M_1, \dots, M_{n-1} M_n$, respectively. If we call N_n and $X_1, \dots, X_n$ the number of intersections of $M_0 M_1 \dots M_n$ and $M_0 M_1, \dots,$

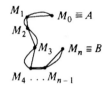

Figure 20. A polygonal approximation of the piece of wire.

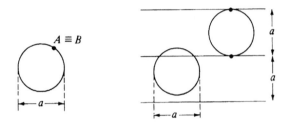

Figure 21. A particular case of Buffon's problem.

$M_{n-1} M_n$, respectively, with the grid, then since $N_n = X_1 + \cdots + X_n$,

$$E[N_n] = E[X_1] + \cdots + E[X_n]$$

and therefore, from the previous results

$$E[N_n] = k l_n \tag{111}$$

where l_n is the total length of the broken line $M_0 M_1 \ldots M_n$. We will call line $\overset{\frown}{AB}$ *regular* if there exists a sequence of broken lines indexed by $n \geqslant 1$ and such that, for all ω, $\lim_{n \uparrow \infty} N_n(\omega) = N(\omega)$ and $N_n(\omega)$ remains bounded, and $\lim_{n \uparrow \infty} l_n = l$. Under these circumstances, $\lim_{n \uparrow \infty} E[N_n] = E[N]$ (Lebesgue's dominated convergence theorem; see Section 4.2), and therefore

$$E[N] = kl. \tag{112}$$

Step 3. It now remains to determine the multiplicative constant k. For this we consider a special line $\overset{\frown}{AB}$ consisting of a circle of diameter a (Fig. 21). In this case N is not random; it is always 2. Therefore, $E[N] = 2$. Also $l = \pi a$, so that by Eq. (112), $2 = k\pi a$, i.e., $k = 2/\pi a$. Finally,

$$E[N] = \frac{2}{\pi} \frac{l}{a}, \tag{113}$$

a surprisingly simple result.

SOLUTIONS FOR CHAPTER 3

E1. By the change of variables $(x - m)/\sigma = y$, we just have to show that

$$\frac{1}{\sqrt{2\pi}} \int_{-\infty}^{+\infty} e^{-(1/2)y^2} \, dy = 1.$$

Now, by Fubini's theorem,

$$\left(\int_{-\infty}^{+\infty} e^{-(1/2)y^2}\, dy \right)^2 = \left(\int_{-\infty}^{+\infty} e^{-(1/2)x^2}\, dx \right) \left(\int_{-\infty}^{+\infty} e^{-(1/2)y^2}\, dy \right)$$

$$= \int_{-\infty}^{+\infty} \int_{-\infty}^{+\infty} e^{-(1/2)(x^2+y^2)}\, dx\, dy.$$

Passing from Cartesian coordinates to polar coordinates, and using Fubini again, we obtain

$$\int_{-\infty}^{+\infty} \int_{-\infty}^{+\infty} e^{-(1/2)(x^2+y^2)}\, dx\, dy = \int_0^{2\pi} \int_0^{\infty} \rho e^{-(1/2)\rho^2}\, d\rho\, d\theta$$

$$= \left(\int_0^{2\pi} d\theta \right) \left(\int_0^{\infty} \rho e^{-(1/2)\rho^2}\, d\rho \right)$$

$$= 2\pi \int_0^{\infty} e^{-u}\, du = 2\pi.$$

Hence, $\int_{-\infty}^{+\infty} e^{-(1/2)y^2}\, dy = \sqrt{2\pi}$, qed.

E2. $$P(Y \leqslant y) = P(\sigma X + m \leqslant y) = P\left(X \leqslant \frac{y-m}{\sigma} \right)$$

$$= \frac{1}{\sqrt{2\pi}} \int_{-\infty}^{(y-m)/\sigma} e^{-(1/2)x^2}\, dx = \frac{1}{\sigma\sqrt{2\pi}} \int_{-\infty}^{y} e^{-(1/2)[(z-m)/\sigma]^2}\, dz$$

$$\left(\text{change of variables } x = \frac{z-m}{\sigma} \right).$$

$$P(X \leqslant x) = P\left(\frac{Y-m}{\sigma} \leqslant x \right) = P(Y \leqslant \sigma x + m)$$

$$= \frac{1}{\sigma\sqrt{2\pi}} \int_{-\infty}^{\sigma x+m} e^{-(1/2)[(y-m)/\sigma]^2}\, dy = \frac{1}{\sqrt{2\pi}} \int_{-\infty}^{x} e^{-(1/2)z^2}\, dz$$

$$\left(\text{change of variables } \frac{y-m}{\sigma} = z \right).$$

E3. $$\int_{-\infty}^{+\infty} f(x)\, dx = \frac{\beta^\alpha}{\Gamma(\alpha)} \int_0^{\infty} x^{\alpha-1} e^{-\beta x}\, dx = \frac{1}{\Gamma(\alpha)} \int_0^{\infty} y^{\alpha-1} e^{-y}\, dy = \frac{\Gamma(\alpha)}{\Gamma(\alpha)} = 1$$

where the second equality has been obtained with the change of variables $y = \beta x$.

E4. $$m_x = \int_{-\infty}^{+\infty} x f(x)\, dx = \frac{\beta^\alpha}{\Gamma(\alpha)} \int_0^{\infty} x^\alpha e^{-\beta x}\, dx = \frac{1}{\beta\Gamma(\alpha)} \int_0^{\infty} y^\alpha e^{-y}\, dy$$

$$= \frac{1}{\beta} \frac{\Gamma(\alpha+1)}{\Gamma(\alpha)} = \frac{\alpha}{\beta}$$

where identity (10) has been used. Similarly,

$$\int_{-\infty}^{+\infty} x^2 f(x)\, dx = \frac{1}{\beta^2} \frac{\Gamma(\alpha+2)}{\Gamma(\alpha)} = \frac{\alpha(\alpha+1)}{\beta^2},$$

so that

$$\sigma_x^2 = \frac{\alpha(\alpha + 1)}{\beta^2} - \frac{\alpha^2}{\beta^2} = \frac{\alpha}{\beta^2}.$$

E5. Case $X \sim \mathcal{U}([a, b])$:

$$m = \frac{1}{b - a} \int_a^b x \cdot 1 \, dx = \frac{1}{b - a} \frac{b^2 - a^2}{2} = \frac{a + b}{2}.$$

Using Eq. (14)

$$\sigma^2 = \frac{1}{b - a} \int_a^b x^2 \, dx - m^2$$

$$= \frac{1}{b - a} \frac{b^3 - a^3}{3} - \left(\frac{a + b}{2}\right)^2 = \frac{(a - b)^2}{12}.$$

Case $X \sim \mathscr{E}(\lambda)$:

$$m = \int_0^\infty x \lambda e^{-\lambda x} \, dx = \text{(integration by parts)} \int_0^\infty e^{-\lambda x} \, dx - (x e^{-\lambda x})_0^\infty = \frac{1}{\lambda} - 0 = \frac{1}{\lambda}.$$

Using Eq. (14)

$$\sigma^2 = \int_0^\infty x^2 \lambda e^{-\lambda x} \, dx - m^2$$

$$= \text{(integration by parts)} - (x^2 e^{-\lambda x})_0^\infty + \int_0^\infty 2\lambda x e^{-\lambda x} - m^2$$

$$= 0 + 2m - m^2 = \frac{2\lambda - 1}{\lambda^2}.$$

Case $X \sim \mathcal{N}(0, 1)$:

$$m = \frac{1}{\sqrt{2\pi}} \int_{-\infty}^{+\infty} x e^{-(1/2)x^2} \, dx = 0 \qquad \text{(by symmetry)}.$$

$$\sigma^2 = \frac{1}{\sqrt{2\pi}} \int_{-\infty}^{+\infty} x^2 e^{-(1/2)x^2} \, dx$$

$$= \text{(integration by parts)} - \frac{1}{\sqrt{2\pi}} (x e^{-(1/2)x^2})_{-\infty}^{+\infty} + \frac{1}{\sqrt{2\pi}} \int_{-\infty}^{+\infty} e^{-(1/2)x^2} \, dx$$

$$= 0 + \frac{1}{\sqrt{2\pi}} \int_{-\infty}^{+\infty} e^{-(1/2)x^2} \, dx = 1.$$

E6. $X \sim \mathscr{E}(\lambda)$: $\phi_X(u) = \int_0^\infty e^{iux} \lambda e^{-\lambda x} \, dx = \int_0^\infty \lambda e^{-(\lambda - iu)x} \, dx.$

Formally,

$$\int_0^\infty \lambda e^{-(\lambda - iu)x} \, dx = -\frac{\lambda}{\lambda - iu} (e^{-(\lambda - iu)x})_0^\infty = \frac{\lambda}{\lambda - iu}.$$

A rigorous computation involves contour integration in the complex plane.

$X \sim \mathcal{N}(0, 1)$:

$$\phi_X(u) = \int_{-\infty}^{\infty} e^{iux} \frac{1}{\sqrt{2\pi}} e^{-(1/2)x^2} dx = \frac{1}{\sqrt{2\pi}} \int_{-\infty}^{+\infty} e^{-(1/2)(x^2 - 2iux)} dx$$

$$= e^{-u^2/2} \frac{1}{\sqrt{2\pi}} \int_{-\infty}^{+\infty} e^{-(1/2)(x - iu)^2} dx.$$

Formally,

$$\frac{1}{\sqrt{2\pi}} \int_{-\infty}^{+\infty} e^{-(1/2)(x - iu)^2} dx = \frac{1}{\sqrt{2\pi}} \int_{-\infty}^{+\infty} e^{-y^2/2} dy = 1.$$

Here again, contour integration can be used to justify this computation.

E7.
$$\frac{\beta^\alpha}{\Gamma(\alpha)} \int_0^\infty x^{\alpha-1} e^{-\beta x} e^{iux} dx = \frac{\beta^\alpha}{\Gamma(\alpha)} \int_0^\infty x^{\alpha-1} e^{-(\beta - iu)x} dx$$

$$= \frac{\beta^\alpha}{(\beta - iu)^\alpha \Gamma(\alpha)} \int_0^{+\infty} y^{\alpha-1} e^{-y} dy = \frac{\beta^\alpha}{(\beta - iu)^\alpha}$$

where the second equality has been obtained after a purely formal change of variable $y = (\beta - iu)x$, which can be fully justified by contour integration in the complex plane.

E8. From E2, $X = \sigma Y + m$ where $Y \sim \mathcal{N}(0, 1)$. Therefore, by Eq. (35), $\phi_X(u) = e^{ium}\phi_Y(\sigma^2 u^2)$. And the result follows by Eq. (33).

E9.
$$\int_{-\infty}^{+\infty} e^{-a|x|} e^{iux} dx = \int_0^\infty e^{x(-a + iu)} dx + \int_{-\infty}^0 e^{x(a + iu)} dx$$

$$= \text{(formally)} \left(\frac{1}{-a + iu} e^{x(-a + iu)} \right)_0^\infty + \left(\frac{1}{a + iu} e^{x(a + iu)} \right)_{-\infty}^0$$

$$= \frac{1}{a - iu} + \frac{1}{a + iu} = \frac{2a}{a^2 + u^2}.$$

In particular, $2/(1 + u^2)$ is the Fourier transform of $e^{-|x|}$. By the inversion formula,

$$e^{-|x|} = \frac{1}{\pi} \int_{-\infty}^{+\infty} e^{-iux} \frac{1}{1 + u^2} du = \frac{1}{\pi} \int_{-\infty}^{+\infty} e^{iux} \frac{1}{1 + u^2} du.$$

Therefore, exchanging the roles of x and u,

$$\int_{-\infty}^{+\infty} e^{iux} \left(\frac{1}{\pi} \frac{1}{1 + x^2} \right) dx = e^{-|u|}.$$

E10. $F(x, y) = P\ (X \leq x, Y \leq y)$ is the probability of the event $A = \{\omega | X(\omega) \leq x, Y(\omega) \leq y\}$. Reasoning as in Example 6 of Chapter 1, we see that

$$F(x, y) = \begin{cases} 1 & \text{if } x \geq 1 \text{ and } y \geq 1 \\ x & \text{if } 0 \leq x \leq 1 \text{ and } y \geq 1 \\ y & \text{if } x \geq 1 \text{ and } 0 \leq y \leq 1 \\ xy & \text{if } 0 \leq x \leq 1 \text{ and } 0 \leq y \leq 1 \\ 0 & \text{if either } x < 0 \text{ or } y < 0. \end{cases}$$

Taking

$$f(x, y) = \begin{cases} 1 & \text{if } 0 \leqslant x \leqslant 1 \text{ and } 0 \leqslant y \leqslant 1 \\ 0 & \text{otherwise,} \end{cases}$$

[more concisely, $f(x, y) = 1_{[0,1] \times [0,1]}(x, y)$], we see that

$$F(x, y) = \int_{-\infty}^{x} \int_{-\infty}^{y} f(u, v)\, du\, dv.$$

E11.

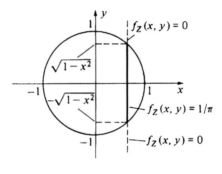

$$f_Z(x, y) = \begin{cases} \dfrac{1}{\pi} & \text{if } x^2 + y^2 \leqslant 1 \\[2mm] 0 & \text{otherwise.} \end{cases}$$

If $|x| \leqslant 1$,

$$f_X(x) = \int_{-\infty}^{+\infty} f_Z(x, y)\, dy = \frac{1}{\pi} \int_{-\sqrt{1-x^2}}^{+\sqrt{1-x^2}} dy = \frac{2}{\pi} \sqrt{1 - x^2}.$$

If $|x| > 1$, $f_X(x) = 0$.

For $0 \leqslant u \leqslant 1$ and $0 \leqslant \theta \leqslant 2\pi$,

$$P(U \leqslant u, \Theta \leqslant \theta) = \frac{\text{shaded area}}{\pi} = \frac{1}{\pi} \pi u^2 \frac{\theta}{2\pi} = \frac{1}{2\pi} u^2 \theta,$$

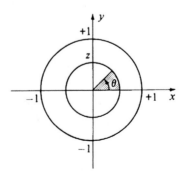

and therefore, in the range of values of (U, Θ), i.e., for $(u, \theta) \in [0, 1] \times [0, 2\pi]$,

$$f_{(U, \Theta)}(u, \theta) = \frac{1}{\pi} u.$$

E12. $m_X = \frac{1}{2\pi} \int_0^{2\pi} \sin \theta \, d\theta = 0, \, m_Y = \frac{1}{2\pi} \int_0^{2\pi} \cos \theta \, d\theta = 0.$

$$\sigma_{XY} = E[(X - m_X)(Y - m_Y)] = E[XY] = E[\cos \Theta \sin \Theta] = E\left[\frac{1}{2} \sin 2\Theta\right]$$

$$= \frac{1}{2\pi} \frac{1}{2} \int_0^{2\pi} \sin 2\theta \, d\theta = 0.$$

E13. From Exercise E11, we know that $m_X = m_Y = 0$ since the densities of X and Y are symmetric around the origin. Therefore,

$$\sigma_{XY} = E[XY] = E[U \cos \Theta U \sin \Theta]$$

$$= E\left[\frac{1}{2} U^2 \sin 2\Theta\right] = \int_0^1 \int_0^{2\pi} \frac{1}{2} u^2 \sin 2\theta \frac{u}{\pi} \, du \, d\theta$$

$$= \frac{1}{2\pi} \int_0^1 u^3 \, du \int_0^{2\pi} \sin 2\theta \, d\theta = 0.$$

X and Y are therefore uncorrelated.

E14. Since $(aX + b) - m_{aX+b} = a(X - m_X)$ and $(cY + d) - m_{cY+d} = c(Y - m_Y)$, we see that if $ac \neq 0$,

$$\rho_{aX+b, cY+d} = \frac{\sigma_{aX, cY}}{\sigma_{aX} \sigma_{cY}} = \frac{ac\sigma_{XY}}{|a|\sigma_X |c|\sigma_Y} = \text{sgn}(ac)\rho_{XY}.$$

If $ac = 0$, the result is trivial since by convention $\rho_{XY} = 0$ when $\sigma_X = 0$.

E16. $\phi_X(0, \ldots, 0, \lambda_j u, 0, \ldots, 0, \lambda_k u, 0, \ldots, 0)$

$$= E[e^{i(\lambda_j u)X_j + i(\lambda_k u)X_k}] = E[e^{iu(\lambda_j X_j + \lambda_k X_k)}].$$

E17. $\phi_X(u) = \left(1 - \frac{i}{\lambda} u\right)^{-1}, \, \frac{\partial^n \phi_X}{\partial u^n}(u)$

$$= \left(-\frac{i}{\lambda}\right)^n (-1)(-2) \ldots (-n)\left(1 - \frac{i}{\lambda} u\right)^{-n-1},$$

$$E[X^n] = \frac{1}{(i)^n} \frac{\partial^n \phi_X}{\partial u^n}(0) = \frac{n!}{\lambda^n}.$$

E18. $f_{U, \Theta}(u, \theta) = \begin{cases} \dfrac{1}{\pi} u = 2u \dfrac{1}{2\pi} = f_1(u)f_2(\theta) & \text{if} \quad u \in [0, 1], \theta \in [0, 2] \\ 0 & \text{otherwise} \end{cases}$

where $f_1(u) = 2u$ verifies $\int_0^1 f_1(u) \, du = (u^2)_0^1 = 1$ and $f_2(\theta) = 1/2\pi$ verifies $\int_0^{2\pi} f_2(\theta) \, d\theta = 1$. Therefore, U and Θ are independent and $f_\Theta(\theta) = (1/2\pi)1_{[0, 2\pi]}(\theta)$, $f_U(u) = 2u 1_{[0, 1]}(u) \, du$. X and Y are not independent since we do not have the factorization $f_{X, Y}(x, y) = f_X(x) f_Y(y)$.

E19.
$$E[e^{iuZ}] = E[e^{iu[(X_1 + \cdots + X_n)/n]}] = E[e^{i(u/n)X_1} \ldots e^{i(u/n)X_n}]$$

$$= \prod_{j=1}^{n} E[e^{i(u/n)X_j}] = \prod_{j=1}^{n} \phi_{X_j}\left(\frac{u}{n}\right) = \prod_{j=1}^{n} e^{-|u/n|} = e^{-|u|}.$$

E20.
$$\phi_{X_1}(u) = \left(1 - \frac{iu}{\beta}\right)^{-\alpha_1}, \phi_{X_2}(u) = \left(1 - \frac{iu}{\beta}\right)^{-\alpha_2}$$

and therefore, since X_1 and X_2 are independent,

$$\phi_{X_1 + X_2}(u) = \phi_{X_1}(u)\phi_{X_2}(u) = \left(1 - \frac{iu}{\beta}\right)^{-(\alpha_1 + \alpha_2)}.$$

E21.
$$\sigma^2 = E[(Z - m)^2] = E\left[\left(\sum_{i=1}^{n} X_i - \sum_{i=1}^{n} m_i\right)^2\right] = E\left\{\left[\sum_{i=1}^{n} (X_i - m_i)\right]^2\right\}$$

$$= E\left[\sum_{i=1}^{n} (X_i - m_i)^2 + 2\sum_{i=1}^{n}\sum_{\substack{j=1 \\ i<j}}^{n} (X_i - m_i)(X_j - m_j)\right]$$

$$= \sum_{i=1}^{n} E[(X_i - m_i)^2] + 2\sum_{i=1}^{n}\sum_{\substack{j=1 \\ i<j}}^{n} E[(X_i - m_i)(X_j - m_j)]$$

$$= \sum_{i=1}^{n} \sigma_i^2 + 0 \qquad \text{since if } i \neq j, \quad E[(X_i - m_i)(X_j - m_j)]$$

$$= E[X_i - m_i]E[X_j - m_j] = 0.$$

E22.

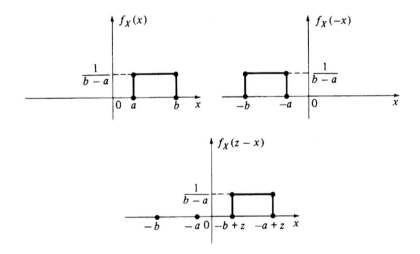

Therefore, $\int_{-\infty}^{+\infty} f_X(z - x)f_Y(x)\,dx = 0$ if $-a + z \leq c$ or $-b + z \geq d$, that is, $f_Z(z) = 0$ if $z \leq a + c$ or $z \geq b + d$. Suppose (without loss of generality because of the symmetric roles of X and Y) that $b - a \leq d - c$. Then from z_1 given by

$c = -b + z_1$ to z_2 given by $d = -a + z_2$, i.e., from $z_1 = b + c$ to $z_2 = a + d$, $\int_{-\infty}^{+\infty} f_X(z - x) f_Y(x) \, dx = f_Z(z)$ remains constant and equal to $(1/b - a)/(1/b - c) \times (d - a) = 1/(d - c)$. Between $a + c$ and $z_1 = b + c$, $f_Z(z)$ increases linearly. Between $z_2 = a + d$ and $b + d$, it decreases linearly.

E23.
$$P(X_1 \in C_1, \ldots, X_n \in C_n) = E[1_{\{X_1 \in C_1, \ldots, X_n \in C_n\}}]$$
$$= E[1_{\{X_1 \in C_1\}} \cdots 1_{\{X_n \in C_n\}}] = E[1_{\{X_1 \in C_1\}}] \cdots E[1_{\{X_n \in C_n\}}]$$
$$= P(X_1 \in C_1) \ldots P(X_n \in C_n).$$

CHAPTER 4

Gauss and Poisson

1. Smooth Change of Variables

1.1. The Method of the Dummy Function

This section is devoted to the computation of the probability density of a random vector $Z = (Z_1, \ldots, Z_p)$ that can be expressed as a sufficiently smooth function of another random vector $X = (X_1, \ldots, X_n)$ of known probability density. The basic tool for doing this is the formula of smooth change of variables in integrals, a result that will be recalled without proof.

Let g be a function from an open set $U \subset \mathbb{R}^n$ into $\mathbb{R}^n$. This function defines a change of variables $x = (x_1, \ldots, x_n) \to y = (y_1, \ldots, y_n)$ by means of $y = g(x)$, or more explicitly,

$$
\begin{cases}
y_1 = g_1(x_1, \ldots, x_n) \\
\ \vdots \\
y_n = g_n(x_1, \ldots, x_n).
\end{cases}
\tag{1}
$$

It will be assumed that the partial derivatives $\partial g_i / \partial x_j$ exist and are continuous in U for all $i, j\ (1 \leqslant i, j \leqslant n)$ and that $|J_g| > 0$ on U, where J_g is the determinant of the Jacobian of g, i.e.,

$$
J_g(x) = \det\left\{ \frac{\partial g_i}{\partial x_j}(x) \right\}.
\tag{2}
$$

With these assumptions the following formula holds:

$$
\int_{g(U)} v(y)\,dy = \int_U v(g(x))|J_g(x)|\,dx,
\tag{3a}
$$

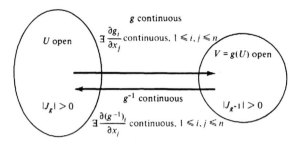

Figure 1. Recapitulation of the conditions on (g, U).

whenever v is a nonnegative or integrable function from $V = g(U) \subset \mathbb{R}^n$ into $\mathbb{R}$.

Under the above assumptions $V = g(U)$ is an open set of $\mathbb{R}^n$ and g admits an inverse $g^{-1}: V \to U$. Moreover $J_{g^{-1}}$ exists and is continuous on V, and $|J_{g^{-1}}| > 0$ on V. Therefore, g^{-1} satisfies the same requirements as g (Fig. 1). Hence, the formula

$$\int_U u(x)\,dx = \int_{g(U)} u(g^{-1}(y))|J_{g^{-1}}(y)|\,dy \qquad (3b)$$

where u is now any nonnegative or integrable function from $U \subset \mathbb{R}^n$ into $\mathbb{R}$.

The formula of change of variables in Eq. (3b) will be applied to the computation of the probability density f_Y of the vector $Y = (Y_1, \ldots, Y_n) = g(X)$, when the probability density f_X of $X = (X_1, \ldots, X_n)$ is null outside U. For this, let us compute, for any nonnegative function h (the dummy function) from $V = g(U) \subset \mathbb{R}^n$ into $\mathbb{R}$, the quantity $E[h(Y)]$:

$$E[h(Y)] = E[h(g(X))] = \int_{\mathbb{R}^n} h(g(x))\,f_X(x)\,dx.$$

Since $f_X = 0$ on $\bar{U}$,

$$E[h(Y)] = \int_U h(g(x))\,f_X(x)\,dx.$$

Now, applying Eq. (3b) to the function $u(x) = h[g(x)]\,f_X(x)$ yields

$$E[h(Y)] = \int_{g(U)} h(y)\,f_X[g^{-1}(y)]|J_{g^{-1}}(y)|\,dy. \qquad (4)$$

This suffices to determine the probability density of Y. Indeed, if for all nonnegative functions $h: \mathbb{R}^n \to \mathbb{R}$, and for some $\phi: \mathbb{R}^n \to \mathbb{R}_+$,

$$E[h(Y)] = \int_{\mathbb{R}^n} h(y)\phi(y)\,dy, \qquad (5)$$

then ϕ is the probability density of Y. To see this, take $h(y) = 1$ if $y \leqslant x$, $h(y) = 0$ if $y > x$. Then Eq. (5) becomes

$$P(Y \leqslant x) = \int_{-\infty}^{x_1} \cdots \int_{-\infty}^{x_n} \phi(y_1, \ldots, y_n)\,dy_1 \ldots dy_n,$$

and therefore by definition, ϕ is the probability density of Y.

Therefore, from Eq. (4),

$$f_Y(y) = f_X[g^{-1}(y)]|J_{g^{-1}}(y)| \qquad \text{on} \quad V = g(U) \tag{6}$$

and $f_Y \equiv 0$ on $\bar{V}$.

1.2. Examples

EXAMPLE 1 (Invertible Affine Transformation). Let Y and X be linked by an affine relation

$$Y = AX + b \tag{7}$$

where A is an $n \times n$ invertible matrix and $b \in \mathbb{R}^n$. Here $g(x) = Ax + b$ and $g^{-1}(y) = A^{-1}(y - b)$, so that

$$|J_{g^{-1}}(y)| = |\det A^{-1}| = \frac{1}{|\det A|}.$$

Therefore,

$$f_Y(y) = \frac{1}{|\det A|} f_X[A^{-1}(y - b)] \tag{8}$$

EXAMPLE 2. Let X_1 and X_2 be two independent random variables uniformly distributed over $[0, 1]$. Find the probability density of X_1/X_2.

Solution. Take $X = (X_1, X_2)$, $Y = (X_1/X_2, X_2)$. The situation is therefore that described above with $U = (0, 1) \times (0, 1)$, $y_1 = g_1(x_1, x_2) = x_1/x_2$, $y_2 = g_2(x_1, x_2) = x_2$. The inverse transformation g^{-1} is described by $x_1 = y_1 y_2$, $x_2 = y_2$, so that

$$|J_{g^{-1}}(y)| = \left| \det \begin{Bmatrix} y_2 & y_1 \\ 0 & 1 \end{Bmatrix} \right| = |y_2|.$$

Also, $V = g(U) = \{(y_1, y_2) | y_1 > 0, 0 < y_2 < \inf(1, 1/y_1)\}$ (Fig. 2). Also,

$$f_X(x_1, x_2) = \begin{cases} 1 & \text{if } x \in U \\ 0 & \text{if } x \notin U. \end{cases}$$

Therefore, by Eq. (6),

$$f_Y(y_1, y_2) = \begin{cases} y_2 & \text{if } y_1 > 0, \qquad 0 < y_2 < \inf\left(1, \dfrac{1}{y_1}\right) \\ \\ 0 & \text{otherwise.} \end{cases}$$

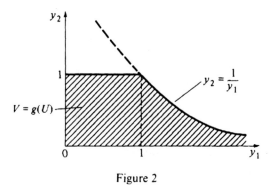

Figure 2

But the problem was to find the probability density f_{Y_1} of $Y_1 = X_1/X_2$. To find it, f_Y must be integrated with respect to its second argument:

$$f_{Y_1}(y_1) = \int_{\mathbb{R}} f_Y(y_1, y_2)\, dy_2 = \int_0^{\inf(1, 1/y_1)} y_2\, dy_2.$$

Finally,

$$f_{Y_1}(y_1) = \frac{1}{2}\left(\inf\left(1, \frac{1}{y_1}\right)\right)^2 \qquad (y_1 > 0).$$

A Word of Caution. One should be aware that on many occasions, the computation of the probability density of $Y = g(X)$ can be done "by hand" without resorting to a formula, as in the exercises that follow.

E1 Exercise. Let X be a real random variable with a distribution function F. Show that $Y = F(X)$ is a random variable uniformly distributed over $[0, 1]$.

E2 Exercise. Show that if X is a Gaussian random variable with mean 0 and variance 1 $[X \sim \mathcal{N}(0, 1)]$, then $Y = X^2$ has the probability density

$$f_Y(y) = \frac{1}{\sqrt{2\pi}} y^{-1/2} e^{-y/2} \qquad (y > 0).$$

E3 Exercise. Show that if $X_1, \ldots, X_n$ are independent Gaussian random variables with mean 0 and variance 1, $Y = X_1^2 + \cdots + X_n^2$ is a chi-square random variable of size n. (*Hint:* Use Exercise E20 of Chapter 3 and E2 above.)

We will have opportunities to apply the general formula of smooth change of random variables in the next sections.

Important Remark. Under the above stated conditions for g (especially $|J_g| > 0$ on the domain of definition U), g is a 1–1 mapping from U onto $V = g(U)$.

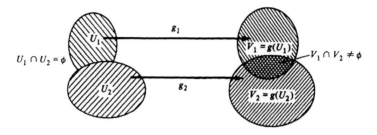

Figure 3. Noninvertible case.

There are cases of pratical interest where the transformation g does not admit an inverse but where its domain U can be decomposed into disjoint open sets (say 2): $U = U_1 + U_2$, such that the restrictions of g to U_1 and U_2, respectively g_1 and g_2, satisfy the conditions of smoothness and of injectivity of the beginning of this section (see Fig. 3). In this case, the same method as above applies, but one must now dissociate the integral:

$$\int_U h(g(x))f_X(x)\,dx = \int_{U_1} h(g_1(x))f_X(x)\,dx + \int_{U_2} h(g_2(x))f_X(x)\,dx$$

and apply the formula of change of variables to each part separately. This gives

$$E[h(Y)] = \int_{g_1(U_1)} h(y)f_X(g_1^{-1}(y))|J_{g_1^{-1}}(y)|\,dy + \int_{g_2(U_2)} h(y)\,f_X(g_2^{-1}(y))|J_{g_2^{-1}}(y)|\,dy$$

and therefore,

$$f_Y(y) = f_X(g_1^{-1}(y))|J_{g_1^{-1}}(y)|1_{g_1(U_1)}(y) + f_X(g_2^{-1}(y))|J_{g_2^{-1}}(y)|1_{g_2(U_2)}(y).$$

2. Gaussian Vectors

2.1. Characteristic Function of Gaussian Vectors

A slight extension of the definition of a Gaussian random variable will be necessary for our purpose, namely: a random variable is said to be Gaussian if its characteristic function is

$$\phi(u) = e^{imu-(1/2)\sigma^2 u^2} \tag{9}$$

where $m \in \mathbb{R}$ and $\sigma \in \mathbb{R}_+$. If $\sigma > 0$, this definition corresponds to the former definition of Chapter 2, section 1.1, and X admits the probability density

$$f(x) = \frac{1}{\sigma\sqrt{2\pi}} e^{-(1/2)[(x-m)^2/\sigma^2]}. \tag{10}$$

If $\sigma = 0$, then X is almost surely a constant, more precisely $X = m$, P-as [see Eq. (28) of Chapter 3]. A Gaussian random vector is defined as follows:

D1 Definition. Let $X = (X_1, \ldots, X_n)$ be a random vector of dimension n. It is said to be Gaussian if and only if the affine combination $a_0 + \sum_{i=1}^{n} a_j X_j$ is a Gaussian random variable, for all $a_0, \ldots, a_n \in \mathbb{R}$.

The characteristic function of such a vector is

$$\boxed{\phi_X(u) = e^{iu'm_X - (1/2)u'\Gamma_X u}} \qquad [u = (u_1, \ldots, u_n)' \in \mathbb{R}^n] \qquad (11)$$

where m_X and Γ_X are the mean vector and the covariance matrix of X, respectively. To see this, it suffices to apply the definition of ϕ_X,

$$\phi_X(u) = E[e^{iu'X}] = E[e^{i\left(\sum_{j=1}^{n} u_j X_j\right)}],$$

and to observe on this expression that $\phi_X(u) = \phi_Y(1)$ where ϕ_Y is the characteristic function of $Y = \sum_{j=1}^{n} u_j X_j$. But by definition, Y is a Gaussian random variable, and therefore

$$\phi_Y(v) = e^{ivm_Y - (1/2)\sigma_Y^2 v^2},$$

where m_Y and σ_Y^2 are the mean and variance of Y, respectively. Now,

$$m_Y = E\left[\sum_{j=1}^{n} u_j X_j\right] = \sum_{j=1}^{n} u_j m_{X_j} = u'm_X,$$

and, remembering a computation in Section 2.2 of Chapter 3,

$$\sigma_Y^2 = \text{var}\left(\sum_{j=1}^{n} u_j X_j\right) = \sum_{i=1}^{n} \sum_{j=1}^{n} \sigma_{ij} u_i u_j = u'\Gamma_X u,$$

where $\{\sigma_{ij}\} = \Gamma_X$ is the covariance matrix of X. Combination of the above computations yields Eq. (11).

A standard n-dimensional Gaussian vector is a n-dimensional Gaussian vector with mean $0 \in \mathbb{R}^n$ and covariance matrix I, the $(n \times n)$ identity matrix. Therefore, if X is a standard n-dimensional Gaussian random vector

$$\phi_X(u_1, \ldots, u_n) = \prod_{j=1}^{n} e^{-u_j^2/2}.$$

Uncorrelated Jointly Gaussian Random Vectors.

T2 Theorem. *Let $X = (Y_1, \ldots, Y_k, Z_1, \ldots, Z_l)'$ be a n-dimensional Gaussian random vector with $n = k + l$. The two random vectors $Y = (Y_1, \ldots, Y_k)'$ and $Z = (Z_1, \ldots, Z_l)'$ are independent if and only if they are uncorrelated.*

Comments. In view of Definition D1, it is clear that Y and Z, being subvectors of a Gaussian vector X, are also Gaussian vectors. A common mistake is to interpret Theorem T2 as follows: if two Gaussian vectors Y and Z are uncorrelated, they are independent. This is *wrong* in general, as Exercise E4 will show. In the statement of Theorem T2, it is required that the "big" vector X obtained by concatenation of Y and Z be also Gaussian. Another way of stating this is: Z and Y are *jointly Gaussian*.

E4 Exercise. Let Y and U be two independent random variables, with $Y \sim \mathcal{N}(0,1)$ and $P(U = -1) = P(U = 1) = \frac{1}{2}$. Let $Z = UY$. Show that $Z \sim \mathcal{N}(0,1)$ and that Y and Z are uncorrelated. Show that Y and Z are *not* independent.

PROOF OF THEOREM T2. Recall the notations of Section 2.2 of Chapter 3, in particular

$$\Sigma_{XY} = E[(X - m_X)(Y - m_Y)']$$

so that

$$\Sigma_{XX} = \begin{pmatrix} \Sigma_{YY} & \Sigma_{YZ} \\ \Sigma_{ZY} & \Sigma_{ZZ} \end{pmatrix},$$

that is,

$$\Gamma_X = \begin{pmatrix} \Gamma_Y & 0 \\ 0 & \Gamma_Z \end{pmatrix}$$

since $\Sigma_{YY} = \Gamma_Y$, $\Sigma_{ZZ} = \Gamma_Z$, $\Sigma_{XX} = \Gamma_X$ and $\Sigma_{XY} = 0$, $\Sigma_{YX} = 0$, according to the hypothesis of uncorrelation. Hence, for all $u = (v', w')' = (v_1, \ldots, v_k, w_1, \ldots, w_l)' \in \mathbb{R}^n$,

$$\phi_X(v_1, \ldots, v_k, w_1, \ldots, w_l) = e^{i(v'm_Y + w'm_Z) - (1/2)(v'\Gamma_Y v + w'\Gamma_Z w)},$$

that is,

$$\phi_{Y,Z}(v_1, \ldots, v_k; w_1, \ldots, w_l) = \phi_Y(v_1, \ldots, v_k)\phi_Z(w_1, \ldots, w_l).$$

But this is a necessary and sufficient condition of independence of Y and Z (Chapter 3, Section 3.2).   □

An immediate corollary of Theorem T2 is that an n-dimensional random vector $X = (X_1, \ldots, X_n)$ is a standard Gaussian vector if and only if the random variables $X_1, \ldots, X_n$ are iid and $X_i \sim \mathcal{N}(0,1)$.

2.2. Probability Density of a Nondegenerate Gaussian Vector

A Gaussian vector X is said to be *nondegenerate* if its covariance matrix Γ_X is positive. In other words, it is nonnegative, that is,

$$u'\Gamma_X u \geqslant 0 \qquad (u \in \mathbb{R}^n), \tag{12a}$$

and moreover (positivity),

$$u'\Gamma_X u = 0 \Rightarrow u = 0. \tag{12b}$$

Recall that property (12a) is shared by all covariance matrices. Also recall that any covariance matrix is symmetric. From Algebra it is known that a strictly positive symmetric matrix Γ is invertible and can be put into the form

$$\Gamma = A \cdot A' \tag{13}$$

where A is an invertible matrix of the same dimension as Γ. Note that such a decomposition is not unique.

Now let X be a *nondegenerate* Gaussian vector with mean m and covariance matrix Γ. It then admits the probability density

$$\boxed{f_X(x) = \frac{1}{(2\pi)^{n/2}(\det \Gamma)^{1/2}} e^{-(1/2)(x-m)'\Gamma^{-1}(x-m)}}. \tag{14}$$

PROOF OF EQ. (14). Consider the vector

$$Z = A^{-1} \cdot (X - m) \tag{15}$$

where A is any $n \times n$ invertible matrix such that $\Gamma = A \cdot A'$. Clearly Z is a Gaussian vector. Indeed, any affine combination of its components is also an affine combination of the components of X and is therefore a Gaussian random variable. We have

$$\begin{cases} m_Z = 0 \\ \Gamma_Z = A^{-1}\Gamma(A^{-1})' = A^{-1}AA'(A^{-1})' = I \end{cases}$$

where I is the $n \times n$ identity matrix. Therefore,

$$\phi_Z(u_1, \ldots, u_n) = \prod_{j=1}^{n} e^{-(1/2)u_j^2}.$$

In view of Eq. (71) of Chapter 3,

$$\phi_{Z_j}(u_j) = \phi_Z(0, \ldots, 0, u_j, 0, \ldots, 0) = e^{-(1/2)u_j^2},$$

that is, $Z_j \sim \mathcal{N}(0,1)$. Moreover, the random variables Z_j $(1 \leqslant j \leqslant n)$ are independent since

$$\phi_Z(u_1, \ldots, u_n) = \prod_{j=1}^{n} \phi_{Z_j}(u_j).$$

The density f of Z is therefore

$$f_Z(z_1, \ldots, z_n) = \frac{1}{(2\pi)^{n/2}} e^{-(1/2)\sum_{j=1}^{n} z_j^2}.$$

The density of X is obtained by Eq. (8):

$$f_X(x) = \frac{1}{(2\pi)^{n/2}} e^{-(1/2)[A^{-1}(x-m)]'A^{-1}(x-m)} |\det A^{-1}|.$$

The result follows by observing that $(A^{-1}(x - m))'A^{-1}(x - m) = (x - m)'$ $\Gamma^{-1}(x - m)$ and $\det A^{-1} = (\det \Gamma)^{-1/2}$. $\qquad\qquad\qquad\qquad\square$

E5 Exercise (Two-Dimensional Gaussian Vector). Let $X = (X_1, X_2)$ be a two dimensional Gaussian vector with mean $m = 0 \in \mathbb{R}^2$ and with the covariance matrix

$$\Gamma = \begin{pmatrix} \sigma_1^2 & \sigma_{12} \\ \sigma_{21} & \sigma_2^2 \end{pmatrix}.$$

What is the relation between σ_{12} and σ_{21}? Give a necessary and sufficient condition for Γ to be nondegenerate in terms of the correlation coefficient ρ. Find a 2×2 invertible matrix A such that $Z = A^{-1}X$ is a standard two-dimensional Gaussian vector, i.e., $Z = (Z_1, Z_2)$, Z_1 independent of Z_2, and Z_1, $Z_2 \sim \mathcal{N}(0, 1)$.

Remark. Let $f(x_1, x_2)$ be the probability density of the two-dimensional Gaussian vector of Exercise E5. The shape of the equidensity lines $f(x_1, x_2) = \lambda$ is given by Fig. 4.

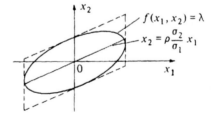

Figure 4. An ellipse of equidensity of a 2-dimensional Gaussian p.d.f.

E6 Exercise (The Degenerate Case). Consider an n-dimensional Gaussian vector X with mean $m = 0$ and a possibly degenerate covariance matrix Γ. Use the decompositions (67) and (68) of Section 2.2. Chapter 3 to prove that if the rank of Γ is $r \leqslant n$, there exists a r-dimensional standard Gaussian vector Y and an $(n \times r)$ matrix D of rank r, such that $X = DY$. Thus, a degenerate n-dimensional vector X lies in a hyperplane π of $\mathbb{R}^n$. The dimension of π is equal to the rank of the covariance matrix.

2.3. Moments of a Centered Gaussian Vector

We shall now give without proof two useful formulas concerning the moments of a centered (0-mean) n-dimensional Gaussian vector $X = (X_1, \ldots, X_n)$ with the covariance matrix $\Gamma = \{\sigma_{ij}\}$. First, we have

$$E[X_{i_1} X_{i_2} \ldots X_{i_{2k}}] = \sum_{\substack{(j_1,\ldots,j_{2k}) \\ j_1 < j_2,\ldots,j_{2k-1} < j_{2k}}} \sigma_{j_1 j_2} \sigma_{j_3 j_4} \cdots \sigma_{j_{2k-1} j_{2k}} \quad , \qquad (16)$$

where the summation extends over all permutations $(j_1,\ldots,j_{2k})$ of $\{i_1,\ldots,i_{2k}\}$ such that $j_1 < j_2, \ldots, j_{2k-1} < j_{2k}$. There are $1 \cdot 3 \cdot 5 \ldots (2k-1)$ terms in the right-hand side of Eq. (16). The indices $i_1, \ldots, i_{2k}$ are in $\{1,\ldots,n\}$ and they may occur with repetitions. A few examples will help understand how to use Eq. (16):

$$E[X_i X_j] = \sigma_{ij} \qquad (\text{not } \sigma_{ij} + \sigma_{ji})$$

$$E[X_i X_j X_k X_l] = \sigma_{ij}\sigma_{kl} + \sigma_{ik}\sigma_{jl} + \sigma_{il}\sigma_{jk}$$

$$E[X_1^{2k}] = 1 \cdot 3 \ldots (2k-1)\sigma_1^{2k}.$$

The computations leading to Eq. (16) are tedious and use Eq. (73) of Chapter 3. They would also show that any odd moment of a centered Gaussian vector is null, that is,

$$E[X_{i_1} \ldots X_{i_{2k+1}}] = 0 \quad , \qquad (17)$$

for all $(i_1,\ldots,i_{2k+1}) \in \{1,2,\ldots,n\}^{2k+1}$.

E7 Exercise. Let $X \sim \mathcal{N}(0,1)$ and let (X_1, X_2) be a two-dimensional standard Gaussian vector. Compute $E[X_1^2 X_2^2]$ and $E[X^4]$.

2.4. Random Variables Related to Gaussian Vectors

Empirical Mean and Empirical Variance of a Gaussian Sample. A Gaussian sample of size n is, by definition, a random vector $X = (X_1, \ldots, X_n)$ where the X_i's are iid with the common distribution $\mathcal{N}(m, \sigma^2)$. Any random variable of the form $f(X_1, \ldots, X_n)$ where $f: \mathbb{R}^n \to \mathbb{R}$, is called a *statistic* of the sample $(X_1, \ldots, X_n)$. The two main statistics are the *empirical mean*

$$\bar{X} = \frac{X_1 + \cdots + X_n}{n} \qquad (18)$$

and the *empirical variance*

$$S^2 = \frac{1}{n-1} \sum_{i=1}^{n} (X_i - \bar{X})^2 \quad . \qquad (19)$$

T3 Theorem. *In a Gaussian sample* $(X_1, \ldots, X_n)$ *of mean m and variance* σ^2, *the empirical mean* $\bar{X}$ *and the empirical variance* S^2 *are independent, and* $[(n-1)/\sigma^2] S^2$ *has a chi-square distribution with* $n - 1$ *degrees of freedom.*

PROOF OF THEOREM T3. Take the case of $m = 0$, $\sigma^2 = 1$. We shall rely on the following lemma of Linear Algebra, which we admit without proof. There exists a $n \times n$ unitary matrix C such that if $x = (x_1, \ldots, x_n)' \in \mathbb{R}^n$ and $y = (y_1, \ldots, y_n)' \in \mathbb{R}^n$ are related by

$$y = Cx, \tag{20}$$

then

$$\begin{cases} y_n = \sqrt{n}\bar{x} \\ y_1^2 + \cdots + y_{n-1}^2 = (n-1)s^2, \end{cases} \tag{21}$$

where

$$\begin{cases} \bar{x} = \dfrac{1}{n}(x_1 + \cdots + x_n) \\ s^2 = \dfrac{1}{n-1} \sum_{i=1}^{n} (x_i - \bar{x})^2. \end{cases} \tag{22}$$

Consider the random vector $Y = CX$ where C is as in the above lemma. It is a Gaussian vector since X is Gaussian. It is, moreover, a standard Gaussian vector since

$$\Gamma_Y = C\Gamma_Y C' = CIC' = CC' = I$$

where we have used the fact that the transpose of a unitary matrix is also its inverse. According to Eqs. (21) and (22),

$$\begin{cases} \bar{X} = Y_n/\sqrt{n} \\ S^2 = \dfrac{1}{n-1}(Y_1^2 + \cdots + Y_{n-1}^2). \end{cases}$$

The independence of $\bar{X}$ and S^2 follows from the independence of Y_n and $(Y_1, \ldots, Y_{n-1})$. The random variable $(n-1)S^2$ is the sum of $n-1$ squared independent standard Gaussian random variables and is therefore chi-square distributed, with $n-1$ degrees of freedom (see E3).

The general case follows from the case $m = 0$, $\sigma^2 = 1$ by considering the standard Gaussian random variables $X_i' = (X_i - m)/\sigma$ $(1 \leqslant i \leqslant n)$ and by observing that $\bar{X}' = (1/n) \sum_{i=1}^{n} X_i' = (\bar{X} - m)/\sigma$ and $(S')^2 = (n-1)^{-1} \sum_{i=1}^{n} (X_i' - \bar{X}')^2 = S^2/\sigma^2$. □

Statistics of Fisher, Snedecor, and Student. Let S_1^2 and S_2^2 be the empirical variances of two *independent* Gaussian samples of size n_1 and n_2, respectively, with the same variance $\sigma_1^2 = \sigma_2^2 = \sigma^2$ and with possibly different means.

The *Fisher statistic* is, by definition, the random variable

$$F = \frac{S_1^2}{S_2^2}. \tag{23}$$

The two following exercises will give the probability density of F.

E8 Exercise. Let X and Y be two independent random variables with

$$X \sim \chi_n^2, \qquad Y \sim \chi_m^2.$$

Compute the probability density of (Z, Y) where $Z = X/Y$. Deduce from this result the probability density of Z.

A *Fisher* random variable of parameters $\alpha > 0$ and $\beta > 0$, is a random variable with the probability density

$$f(x) = \frac{\Gamma\left(\dfrac{\alpha + \beta}{2}\right) \alpha^{\alpha/2} \beta^{\beta/2}}{\Gamma\left(\dfrac{\alpha}{2}\right) \Gamma\left(\dfrac{\beta}{2}\right)} \frac{x^{\alpha/2 - 1}}{(\beta + \alpha x)^{(\alpha + \beta)/2}}. \qquad (x \geqslant 0) \tag{24}$$

E9 Exercise. Show that F given by Eq. (23) is a Fisher random variable with parameters $n_1 - 1$ and $n_2 - 1$.

A *Fisher–Snedecor* random variable of parameters $\alpha > 0$ and $\beta > 0$ is a random variable with the probability density

$$f(x) = \frac{2\Gamma\left(\dfrac{\alpha + \beta}{2}\right)}{\Gamma\left(\dfrac{\alpha}{2}\right) \Gamma\left(\dfrac{\beta}{2}\right)} \frac{x^{\alpha - 1}}{(1 + x^2)^{(\alpha + \beta)/2}}. \qquad (x \geqslant 0) \tag{25}$$

E10 Exercise. For the random variable F being defined by Eq. (23) show that $F^{1/2} = \sqrt{(n_1 - 1)/(n_2 - 1)}(|S_1|/|S_2|)$ is a Fisher–Snedecor random variable with parameters $n_1 - 1$ and $n_2 - 1$.

A *Student* random variable with the parameter $\alpha > 0$ is a random variable with the probability density

$$f(x) = \frac{\Gamma\left(\dfrac{\alpha + 1}{2}\right)}{(\alpha\pi)^{1/2} \Gamma\left(\dfrac{\alpha}{2}\right)} \left(1 + \frac{x^2}{\alpha}\right)^{-(\alpha + 1)/2}. \tag{26}$$

The *Student statistic* T of a Gaussian sample of size $n \geq 2$ defined by

$$T = \frac{n^{-1/2}(\bar{X} - m)}{|S|}.$$ (27)

E11 Exercise. Show that T is a Student random variable with parameter $n - 1$.

Remark. The *Cauchy* density

$$f(x) = \frac{1}{\pi} \frac{1}{1 + x^2}$$ (28)

is obtained by setting $\alpha = 1$ in Eq. (26). Its shape is shown in Fig. 5.

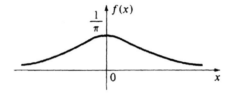

Figure 5. The Cauchy density.

3. Poisson Processes

3.1. Homogeneous Poisson Processes Over the Positive Half Line

Let us consider a sequence of positive real valued random variables $(T_n, n \geq 1)$ that is strictly increasing:

$$0 < T_1 < T_2 < \cdots < T_n < T_{n+1} < \cdots < \infty.$$ (29)

The random variable T_n can be thought of as the arrival time of the nth customer requiring service in a given service system (freeway toll, movie theater, etc.). The random variable T_0 is conventionally taken to be 0. The interarrival sequence associated to $(T_n, n \geq 1)$ is $(S_n, n \geq 1)$ where

$$S_{n+1} = T_{n+1} - T_n \qquad (n \geq 0).$$ (30)

The *counting process* associated with $(T_n, n \geq 1)$ is the family of $\mathbb{N}$-valued random variables $(N(t), t \geq 0)$ defined by

$$N(t) = n \qquad \text{if} \quad T_n \leq t < T_{n+1} \qquad (n \geq 0).$$ (31)

Thus, $N(t)$ is the number of arrivals during the time interval $(0, t]$. Figure 6 summarizes these definitions. The sequence $(T_n, n \geq 1)$ is also called a "point process" over $\mathbb{R}_+$, and T_n is the nth random "point" of the point process.

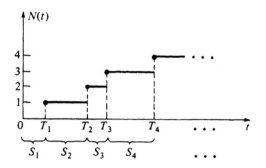

Figure 6. A point process and its associated counting process.

D4 Definition. The point process $(T_n, n \geq 1)$ over R_+ is called a homogeneous Poisson process of intensity $\lambda > 0$ iff its associated counting process $(N(t), t \geq 0)$ verifies the following:

(i) for all $a, b \in \mathbb{R}_+$ such that $a \leq b$, $N(b) - N(a)$ is a Poisson random variable of mean $(b - a)\lambda$ i.e.,

$$P(N(b) - N(a) = k) = e^{-(b-a)\lambda}\frac{(b - a)^k \lambda^k}{k!} \qquad (k \in \mathbb{N}); \qquad (32)$$

(ii) $(N(t), t \geq 0)$ has independent increments, that is, for all $0 \leq t_1 \leq t_2 \leq \cdots \leq t_m$, the random variables $N(t_2) - N(t_1), N(t_3) - N(t_2), \ldots, N(t_m) - N(t_{m-1})$ are independent.

If $(T_n, n \geq 1)$ is a Poisson process with the intensity $\lambda > 0$, the distribution of $T_n (n \geq 1)$ is easily obtained from the following remark:

$$T_n \leq t \Leftrightarrow N(t) \geq n.$$

Therefore, if we denote by F_{T_n} the cumulative distribution function of T_n,

$$F_{T_n}(t) = P(T_n \leq t) = P(N(t) \geq n) = 1 - P(N(t) \leq n - 1),$$

that is, in view of (i) of Definition D4,

$$F_{T_n}(t) = 1 - \sum_{k=0}^{n-1} e^{-\lambda t}\frac{(\lambda t)^k}{k!} \qquad (n \geq 1, t \geq 0). \qquad (33)$$

The probability density f_{T_n} of T_n is obtained by differentiation of F_{T_n}:

$$f_{T_n}(t) = \begin{cases} \lambda e^{-\lambda t}\dfrac{(\lambda t)^{n-1}}{(n - 1)!} & \text{if } t \geq 0. \\ 0 & \text{otherwise} \end{cases} \qquad (34)$$

In particular, for $n = 1$,

$$f_{T_1}(t) = \lambda e^{-\lambda t} \qquad (t \geqslant 0). \tag{35}$$

In otherwords, T_1 is an exponential random variable with mean $1/\lambda$.

If we take a closer look at expression (34), we recognize a familiar density probability: the gamma density with parameters $\alpha = n$ and $\beta = \lambda$ [see Eq. (8) of Chapter 3].

E12 Exercise. Let $X_1, \ldots, X_n$ be iid random variables with $X_i \sim \mathscr{E}(\lambda)$. Show that

$$X_1 + \cdots + X_n \sim \gamma(n, \lambda).$$

We have

$$T_n = S_1 + \cdots + S_n. \tag{36}$$

Exercise E12 and Eq. (36) tell us that everything is "just as if" $S_1, \ldots, S_n$ were iid with $S_i \sim \mathscr{E}(\lambda)$. In fact, this is true, as shown in the following exercises.

E13 Exercise. Let $(T_n, n \geqslant 1)$ be a homogeneous Poisson process over $\mathbb{R}_+$ with intensity $\lambda > 0$. Compute the pd of $(T_1, \ldots, T_k)$ for any $k \in \mathbb{N}_+$. Apply the formula of smooth change of variables to obtain the pd of $(S_1, \ldots, S_k)$ and conclude that $S_1, \ldots, S_k$ are iid with $S_i \sim \mathscr{E}(\lambda)$, as announced.

E14 Exercise. Let $(T_n, n \geqslant 1)$ be a sequence of nonnegative real valued random variables such that $0 < T_1 \leqslant T_2 \leqslant T_3 \leqslant \cdots$ and define for each $t \in \mathbb{R}_+$, $N(t)$ to be the cardinality of the set $\{n \in \mathbb{N}_+ | T_n \in (0, t]\}$. Suppose that (i) and (ii) of Definition D4 are satisfied. Show that for almost all ω, there is no finite accumulation point of the sequence $(T_n(\omega), n \geqslant 1)$, and that $(T_n(\omega), n \geqslant 1)$ is a *strictly* increasing sequence.

Exercise E14 tells us that the strict inequality signs in (29) were superfluous in the definition of a homogeneous Poisson process.

3.2. Nonhomogeneous Poisson Process Over the Positive Half Line

D5 Definition. Let $(T_n, n \geqslant 1)$ be a point process over $\mathbb{R}_+$ and let $\lambda : \mathbb{R}_+ \to \mathbb{R}$ be a nonnegative function such that

$$\int_0^t \lambda(s)\, ds < \infty \qquad (t \geqslant 0). \tag{37}$$

If the counting process $(N(t), t \geqslant 0)$ associated with $(T_n, n \geqslant 1)$ verifies

(i) for all $a, b \in \mathbb{R}_+$ such that $a \leqslant b$, $N(b) - N(a)$ is a Poisson random variable of mean $\int_a^b \lambda(s)\, ds$, and

(ii) $(N(t), t \geqslant 0)$ has independent increments [see (ii) of D4],

then $(T_n, n \geqslant 1)$ is called a nonhomogeneous Poisson process with intensity $\lambda(t)$.

E15 Exercise. What is the probability density of T_n when $(T_n, n \geqslant 1)$ is a Poisson process with intensity $\lambda(t)$? What is the distribution function of T_1? Show that since $T_1 < \infty$, necessarily $\int_0^\infty \lambda(s)\, ds = \infty$.

E16 Exercise (Campbell's Formula). Let $(T_n, n \geqslant 1)$ be a Poisson process with intensity $\lambda(t)$. Let $(X_n, n \geqslant 1)$ be an iid sequence with $P(X_1 \leqslant x) = F(x)$. Suppose that $(T_n, n \geqslant 1)$ and $(X_n, n \geqslant 1)$ are independent. Show that for any nonnegative bounded function $g : \mathbb{R} \times \mathbb{R} \to \mathbb{R}$

$$E\left[\sum_{n=1}^{\infty} g(T_n, X_n) 1_{\{T_n \leqslant t\}} \right] = \int_0^t E[g(s, X_1)] \lambda(s)\, ds. \tag{38}$$

(You may suppose that X_1 admits a pd although the formula is true without this assumption.)

E17 Exercise ($M/G/\infty$ Service System). The notation "$M/G/\infty$" symbolizes a service system in which the arrival times of customers $(T_n, n \geqslant 1)$ form a homogeneous Poisson process with intensity $\lambda > 0$, and the time spent by customer $\#n$ (arriving at T_n) is a random variable X_n. The sequences $(T_n, n \geqslant 1)$ and $(X_n, n \geqslant 1)$ are supposed to be independent, and $(X_n, n \geqslant 1)$ is an iid sequence with $P(X_n \leqslant x) = F(x)$ (Fig. 7).

Let $X(t)$ be the number of customers present in the system at time $t > 0$, with the assumption that the system is empty at time $t = 0$. Compute $E[X(t)]$ using the result of Exercise E16. Using the formula $E[X] = \int_0^\infty [1 - G(x)]\, dx$ valid for any nonnegative random variable X with $P(X \leqslant x) = G(x)$, show that

$$\lim_{t \uparrow \infty} E[X(t)] = \lambda E[X_1].$$

What is the explicit form of $E[X(t)]$ when $X_1 \sim \mathscr{E}(\mu)$?

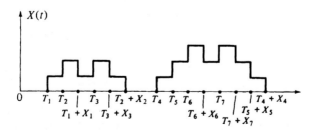

Figure 7. Evolution of the queueing process in a $M/G/\infty$ system.

3.3. Homogeneous Poisson Processes on the Plane

A point process on the plane is a probabilistic model for random points scattered in the plane. It is described by a collection of random variables $\{N(A);\ A \in \mathscr{B}^2\}$ indexed by the borelian sets of $\mathbb{R}^2$, where $N(A)$ is the (random) number of (random) points in A. It will be assumed that if A is bounded, i.e., if it is contained in a rectangle, then $N(A, \omega)$ is finite for P-almost all ω. Also, there are no "multiple points," that is, for all $a \in \mathbb{R}^2$, $N(\{a\}) = 0$ or 1, P-almost surely, where $\{a\}$ is the set consisting of the element a only.

It is then said that "N is a simple point process."

D6 Definition. A homogeneous Poisson process on $\mathbb{R}^2$ with intensity $\lambda > 0$ is a simple point process where in addition:

(α) For any finite sequence $A_1, \ldots, A_n$ of disjoint bounded Borel sets of $\mathbb{R}^2$, the random variables $N(A_1), \ldots, N(A_n)$ are independent.

(β) For any bounded borelian set A of $\mathbb{R}^2$,

$$P(N(A) = k) = e^{-\lambda S(A)} \frac{(\lambda S(A))^k}{k!} \qquad (k \in \mathbb{N}) \tag{39}$$

where $S(A)$ is the area of A.

E18 Exercise. Let N be a homogeneous point process on $\mathbb{R}^2$ with intensity 1, and let $\lambda \colon \mathbb{R}_+ \to \mathbb{R}$ be a nonnegative continuous function such that $\int_0^t \lambda(x)\, dx < \infty$ for all $t \in \mathbb{R}^+$. Define the random variable $N(t)$ for each $t \in \mathbb{R}_+$ as the number of points of the process N in the set $A_t = \{(x, y) \in \mathbb{R}^2 \,|\, x \in (0, t],\ 0 \leqslant y \leqslant \lambda(x)\}$ (Fig. 8). Show that $(N(t), t \in \mathbb{R}_+)$ verifies (i') and (ii') of Definition D5 of a nonhomogeneous Poisson process over the positive half-line.

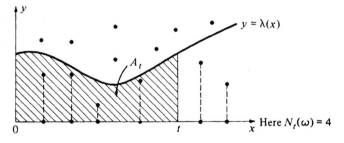

Figure 8. Points below a curve.

The Poisson Approximation. Consider the square S_D centered on 0, with sides parallel to the coordinate axes and of size D (Fig. 9). Let K be, for the time being, a fixed integer, and consider K independent couples of random vari-

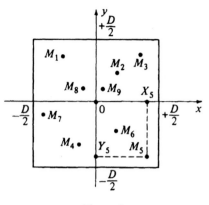

Figure 9

ables (X_i, Y_i), $i = 1, \ldots, K$ such that the point $M_i = (X_i, Y_i)$ is uniformly distributed on S_D, i.e.,

$$P((X_i, Y_i) \in A) = \frac{S(A)}{D^2} \qquad \text{for all} \quad A \in \mathcal{B}^2, A \subset S_D. \qquad (40)$$

This models K points independently thrown at random uniformly onto the square S_D.

Consider now a sequence of disjoint bounded sets $A_1, \ldots, A_n$ of $\mathbb{R}^2$ all contained in S_D. The joint distribution of $N(A_1), \ldots, N(A_n)$ (the respective number of points in $A_1, \ldots, A_n$) is computed as follows:

$$P(N(A_1) = k_1, \ldots, N(A_n) = k_n) = P(N(A_0) = k_0, \ldots, N(A_n) = k_n)$$

where $A_0 = S_D - \sum_{j=1}^{n} A_j$ and $k_0 = K - \sum_{j=1}^{n} k_j$, and since each point M_i has the probability $p_i = [S(A_i)]/D^2$ of falling in A_i, we obtain the multinomial distribution of size (K, n) [see Chapter 2, Section 1.1 and Eq. (11)].

$$P(N(A_1) = k_1, \ldots, N(A_n) = k_n) = \frac{K!}{k_0! \ldots k_n!} p_0^{k_0} \ldots p_n^{k_n}. \qquad (41)$$

E19 Exercise. Show that if D and K tend simultaneously toward ∞, in such a manner that $K/D^2 = \lambda > 0$ for a fixed $\lambda > 0$, then the right-hand side of Eq. (41) tends to

$$\prod_{i=1}^{n} e^{-\lambda S(A_i)} \frac{(\lambda S(A_i))^{k_i}}{k_i!}. \qquad (42)$$

Since Eq. (42) is the distribution of $N(A_1), \ldots, N(A_n)$ when N is a homogeneous Poisson process over $\mathbb{R}^2$ of intensity $\lambda > 0$, we have obtained an intuitive interpretation for the statistical distribution of the points of a homogeneous Poisson process.

The above result can be used as follows: if you have a very large area (say S_D) on which many points (say K) have been thrown independently at random uniformly into this area, and if you want to compute the quantity

$$P(N(A_1) = k_1, \ldots, N(A_n) = k_n)$$

when $k_1 + \cdots + k_n$ is much smaller than K, and when $A_1, \ldots, A_n$ are far from the boundaries of S_D, then you can use the "Poisson approximation" (42).

4. Gaussian Stochastic Processes

4.1. Stochastic Processes and Their Laws

The counting process $(N(t), t \geq 0)$ of a point process $(T_n, n \geq 1)$ is an example of stochastic process, a notion that will now be introduced more formally.

D7 Definition. A (real valued) stochastic process defined on $(\Omega, \mathscr{F}, P)$ is a family $[X(t), t \in \mathbb{T}]$ of (real valued) random variables on $(\Omega, \mathscr{F}, P)$ where the index set $\mathbb{T}$ is an interval of $\mathbb{R}$.

Thus, for each $t \in \mathbb{T}$, $\omega \to X(t, \omega)$ is a random variable. Another way of looking at the function of two arguments $(t, \omega) \to X(t, \omega)$ consists of observing that for each $\omega \in \Omega$, $t \to X(t, \omega)$ is a (real valued) function defined on $\mathbb{T}$. When adopting the latter point of view, a stochastic process is also called a *random function*.

If for P-almost all $\omega \in \Omega$ the function $t \to X(t, \omega)$ is continuous on $\mathbb{T}$, the stochastic process $(X(t), t \in \mathbb{T})$ is said to be *continuous*. Similar notions are available, such as right-continuous stochastic process, increasing stochastic process, etc. For instance, the counting process of a point process over $\mathbb{R}_+$ is a right-continuous increasing stochastic process.

The statistical behavior of a stochastic process $(X(t), t \in \mathbb{T})$ is described by its *law*. To give the law of such a stochastic process consists in giving for all $n \in \mathbb{N}_+$, and all $t_1, \ldots, t_n \in \mathbb{T}$ the distribution of the random vector $(X(t_1), \ldots, X(t_n))$. Recall that the distribution of a random vector is equivalently described by its cumulative distribution function, its characteristic function, or its probability density when such density exists. For instance, the law of the counting process $(N(t), t \in \mathbb{R}_+)$ associated with a homogeneous Poisson process over $\mathbb{R}_+$ with intensity $\lambda > 0$ is described by the data

$$P(N(t_1) = k_1, \ldots, N(t_n) = k_n)$$

$$= e^{-\lambda t_1} \frac{(\lambda t_1)^{k_1}}{k_1!} \frac{(\lambda(t_2 - t_1))^{k_2 - k_1}}{(k_2 - k_1)!} \cdots \frac{(\lambda(t_n - t_{n-1}))^{k_n - k_{n-1}}}{(k_n - k_{n-1})!}$$

for all $t_1, t_2, \ldots, t_n \in \mathbb{R}_+$ such that $t_1 \leqslant t_2 \leqslant \cdots \leqslant t_n$ and for all $k_1, k_2, \ldots, k_n \in \mathbb{N}$ such that $k_1 \leqslant k_2 \leqslant \cdots \leqslant k_n$.

4.2. Gaussian Stochastic Processes

The most famous example of continuous stochastic process is the Wiener process or Brownian motion first studied by Einstein in view of modeling diffusion phenomena.

D8 Definition. A real valued continuous stochastic process $(W(t), t \in \mathbb{R}_+)$ is called a Wiener process if $W(0) = 0$, P-as and if for all $t_1, t_2, \ldots, t_n \in \mathbb{R}_+$ such that $t_1 \leqslant t_2 \leqslant \cdots \leqslant t_n$, the random variables $W(t_2) - W(t_1), \ldots, W(t_n) - W(t_{n-1})$ are Gaussian and independent, with mean 0 and respective variances $t_2 - t_1, \ldots, t_n - t_{n-1}$.

The probability density of $(W(t_1), W(t_2) - W(t_1), \ldots, W(t_n) - W(t_{n-1}))$ where $0 < t_1 < \cdots < t_n$ is therefore

$$\frac{1}{(2\pi)^{n/2}} \exp \left\{ -\frac{1}{2} \left(\frac{x_1^2}{t_1} + \frac{x_2^2}{t_2 - t_1} + \cdots + \frac{x_n^2}{t_n - t_{n-1}} \right) \right\}.$$

E20 Exercise. Show that the probability density of $(W(t_1), \ldots, W(t_n))$ where $0 < t_1 < \cdots < t_n$ is

$$\frac{1}{(2\pi)^{n/2}} \exp \left\{ -\frac{1}{2} \left(\frac{x_1^2}{t_1} + \frac{(x_2 - x_1)^2}{t_2 - t_1} + \cdots + \frac{(x_n - x_{n-1})^2}{t_n - t_{n-1}} \right) \right\}. \tag{43}$$

The Wiener process is a particular case of Gaussian process.

D9 Definition. A real valued stochastic process $(X(t), t \in \mathbb{T})$ is called Gaussian if for all $t_1, \ldots, t_n \in \mathbb{T}$, the random vector $(X(t_1), \ldots, X(t_n))$ is Gaussian.

Therefore, if we define

$$m_X(t) = E[X(t)] \tag{44}$$

$$\Gamma_X(t, s) = E[(X(t) - m_X(t))(X(s) - m_X(s))], \tag{45}$$

the law of a Gaussian process $(X(t), t \in \mathbb{T})$ is given by the characteristic functions

$$E[e^{i \sum_{k=1}^{n} u_k X(t_k)}] = \exp \left\{ i \sum_{k=1}^{n} u_k m_X(t_k) - \frac{1}{2} \sum_{k=1}^{n} \sum_{l=1}^{n} \Gamma_X(t_k, t_l) u_k u_l \right\}. \tag{46}$$

A Gaussian processes is very special in many respects. For instance, its law is entirely described by its mean function $m_X : \mathbb{T} \to \mathbb{R}$ and its *autocorrelation function* $\Gamma_X : \mathbb{T} \times \mathbb{T} \to \mathbb{R}$ given by Eqs. (44) and (45), respectively. For most stochastic processes the data m_X and Γ_X are not sufficient to write down their law.

In Signal theory, Gaussian processes (called "Gaussian signals") are appreciated because if they serve as inputs into a linear filter, the outputs are also Gaussian signals. [The definition of a linear filter and the stability property just mentioned are the object of one of the illustrations of Chapter 5.]

Illustration 7. An Introduction to Bayesian Decision Theory: Tests of Gaussian Hypotheses

Noisy Distance Measurements. Telemetry is the art of measuring the distance between you and inaccessible objects. There are many ways of performing such measurements depending on the available technology and the medium between you and the target. Let us consider sonar telemetry. The basic principle of a distance measurement underwater is the following.

A sonar emits an acoustic pulse and waits for the return of the echo. Dividing the time of return by twice the speed of acoustic waves in the water yields the distance from the emitter (the sonar) to the target. Figure 10 depicts an ideal situation: first, the echo looks exactly like the original pulse (i.e., there is no distorsion), and second, the ambient noise of the sea is null.

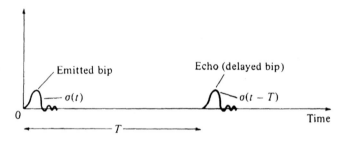

Figure 10. The emitted pulse and its return.

For the time being, it will be assumed that the emitted bip returns to the sonar unaltered (this is grossly unrealistic, but we shall provide means of treating more realistic situations later). However, the ambient noise will be taken into consideration.

Denoting $(\sigma(t), t \in [0, a])$ as the emitted bip and T as the echo return time, the received signal takes the form

$$X(t) = \sigma(t - T) + B(t) \qquad (t \geq 0) \tag{47}$$

where $B(t)$ is for each t a real valued random variable with mean 0. Here we have implicitly assumed that the noise is *additive*.

The received waveform will appear as in Fig. 11. The problem is to decide where the return pulse is. This decision process must be performed automatically. The corresponding signal processing operation (extracting a return time) is performed in several steps. First, as in virtually all signal processors, the received signal is reduced to a form suitable for automatic computations: it must be sampled, i.e., $(X(t), t \geq 0)$ is reduced to a sequence $(X_n, n \geq 1)$ of "samples," where X_n is the sample at time $n\Delta$, $1/\Delta$ being the sample rate:

$$X_n = X(n\Delta) \qquad (n \geq 1).$$

Illustration 7. An Introduction to Bayesian Decision Theory   149

Figure 11. A typical waveform at the receiver.

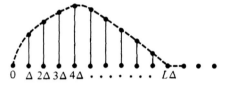

Figure 12. The sampled pulse.

The time extension a of the emitted bip can be so chosen that $a = L\Delta$ for some integer L. The pulse samples $\sigma_n = \sigma(n\Delta)$ are therefore null for $n > L$ (Fig. 12). It will be supposed that the return time T is of the form $i\Delta$ where i is unknown. This hypothesis is innocuous if the energy of the bip contained in an interval of length Δ is small, as is always the case.

Defining $B_n = B(n\Delta)$ for $n \geqslant 1$, the sampled data is therefore

$$
\begin{aligned}
X_1 &= B_1 \\
X_2 &= B_2 \\
&\;\vdots \\
X_{i-1} &= B_{i-1} \\
X_i &= B_i + \sigma_0 \\
X_{i+1} &= B_{i+1} + \sigma_1 \\
&\;\vdots \qquad\qquad\qquad \left.\right\} \text{returning bip} \\
X_{i+L} &= B_{i+L} + \sigma_L \\
X_{i+L+1} &= B_{i+L+1} \\
&\;\vdots \\
X_N &= B_N.
\end{aligned}
$$

Here we stop at N with $N = M + L$, which means that we are interested in targets at a distance such that the return time of the echo is less than $M\Delta$ (Fig. 13).

The mathematical situation after sampling is described by a random vector $X = (X_1, \ldots, X_N)$, which is the sum of another random vector $B = (B_1, \ldots, B_N)$

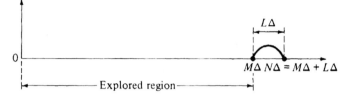

Figure 13

and of a "deterministic" vector $m_i = (0, 0, \ldots, 0, \sigma_0, \ldots, \sigma_L, 0, \ldots, 0)$ where σ_0 is in the ith position. Thus,

$$X = m_i + B. \tag{48}$$

Equality (48) is valid if the return time is $i\Delta$. But the return time is not known, i.e., i is not known. Since it is not known and since any situation can occur a priori, it is wise to consider that the return time is random, or equivalently, that i is the actual value of a random variable Θ with range $\{1, 2, \ldots, M\}$.

The observation vector X is therefore of the form

$$X = m_\Theta + B \tag{49}$$

where $P(\Theta = i) = 1/M$ ($i = 1, \ldots, M$) (the target could be anywhere in the region explored). Our problem is to extract an estimate $\hat{\Theta}$ of Θ based on the observation X and on statistical information concerning the "noise" B. It is often realistic to model the ambient noise $(B(t), t \geqslant 0)$ by assuming that the sampled version of it, i.e., vector B, is a Gaussian vector with mean 0 and covariance matrix $\sigma^2 I$ where I is the $N \times N$ identity matrix. This means that the samples $B_1, \ldots, B_N$ are iid with common distribution $\mathcal{N}(0, \sigma^2)$. Such a noise is called an *additive white Gaussian* noise.

The random elements Θ and B are independent because the position of the target and the ambient sea noise are independent. Therefore we have

$$P(X \in A | \Theta = i) = P(B + m_\Theta \in A | \Theta = i) = P(B + m_i \in A)$$

i.e.,

$$P(X \in A | \Theta = i) = \frac{1}{(2\pi\sigma^2)^{N/2}} \int_A \exp\left\{ -\frac{1}{2\sigma^2} \|x - m_i\|^2 \right\} dx. \tag{50}$$

Recall for later reference, that

$$P(\Theta = i) = \frac{1}{M}. \tag{51}$$

Testing Bayesian Hypotheses. The telemetry problem has now been transformed into a probabilistic problem for which a theory is available, namely the *theory of bayesian tests of hypotheses.* As we are going to present it, the theory of bayesian tests of hypotheses has a range of application extending beyond the telemetry problem, and therefore the objects such as X and Θ will be of a slightly more general form than what is needed to accommodate the special situation of interest to us right now.

Illustration 7. An Introduction to Bayesian Decision Theory   151

The framework is as follows. There are two random elements, X and Θ. The random variable Θ is discrete and takes its values in $\{1, 2, \ldots, M\}$, called the *space of hypotheses* (or "*states of Nature*"). When $\Theta = i$, one says that hypothesis i is in force (or that Nature is in state i). The state of Nature Θ is not directly observed. Instead, a random vector X of dimension N, called the *observation*, is divulged, and on the basis of this information, the statistician (you) must guess in which state nature is. More explicitly, the goddess of chance draws an ω, and as a consequence, $\Theta(\omega)$ and $X(\omega)$ are determined. You are informed about the value of $X(\omega)$ and you must make your guess $\hat{\Theta}(\omega)$ in terms of $X(\omega)$, i.e., $\hat{\Theta}(\omega) = s(X(\omega))$ where s is a function from $\mathbb{R}^n$ into $\{1, \ldots, M\}$.

Of course, you cannot expect to guess correctly all the time, unless X contains all the information about Θ. If $\Theta(\omega) \neq \hat{\Theta}(\omega)$, you make a mistake. Since you are generally required to perform this guess work over and over, the important parameter is the *probability of error*

$$P_E = P(\Theta \neq \hat{\Theta}). \tag{52}$$

Now P_E depends on the guessing strategy, i.e., the mapping $s : \mathbb{R}^n \to \{1, \ldots, M\}$. This dependence will be taken into account in the notation. Thus,

$$\boxed{P_E(s) = P(\Theta \neq s(X))}. \tag{53}$$

One seeks the best strategy, i.e., the strategy s^* that minimizes $P_E(s)$. The best strategy s^* is called *the optimal Bayesian test* of the hypotheses $\{1, \ldots, M\}$ based on the observation of X.

If one is to compute a quantity such as $P(\Theta \neq s(X))$ for a given strategy s, one must know the joint probability law of the pair (Θ, X), which is described in the following terms. There exists a set of probability densities on $\mathbb{R}^n$, $\{f_i(x), 1 \leq i \leq M\}$, and a probability distribution $\pi = (\pi(i), 1 \leq i \leq M)$ over $\{1, \ldots, M\}$ such that

$$P(\Theta = i) = \pi(i) \qquad (1 \leq i \leq M) \tag{54}$$

and

$$P(X \in A | \Theta = i) = \int_A f_i(x)\, dx \qquad (1 \leq i \leq M, A \in \mathscr{B}^n). \tag{55}$$

The function f_i is the pd of observation X when hypothesis i is in force. The law of (Θ, X) is determined by Eqs. (54) and (55) since, by the rule of exhaustive and incompatible causes,

$$P(X \in A, \Theta = i) = \pi(i) \int_A f_i(x)\, dx.$$

In the telemetry example, we have [Eqs. (50) and (51)]

$$\pi(i) = \frac{1}{M}, \qquad f_i(x) = \frac{1}{(2\pi\sigma^2)^{N/2}} \exp\left\{ -\frac{1}{2\sigma^2} \|x - m_i\|^2 \right\}. \tag{56}$$

The computation of $P_E(s)$, the probability of error associated with strategy s, is now feasible. Before performing it, it is useful to observe that any function s from $\mathbb{R}^n$ into $\{1, 2, \ldots, M\}$ is bijectively associated with a partition $\{A_1, \ldots, A_M\}$ of $\mathbb{R}^n$. Indeed, if s is such a function, it suffices to set $A_i = \{x \in \mathbb{R}^n | s(x) = i\}$ to obtain such a partition, and if the partition is given, it suffices to define s by $s(x) = i$ if $x \in A_i$ to obtain a strategy.

We shall work with the representation $\{A_1, \ldots, A_M\}$ of a strategy s. By the theorem of exhaustive and incompatible causes,

$$P(\Theta \neq \hat{\Theta}) = \sum_{i=1}^{M} P(\Theta \neq \hat{\Theta} | \Theta = i) P(\Theta = i) = \sum_{i=1}^{M} \pi(i) P(\hat{\Theta} \neq i | \Theta = i).$$

Since strategy $s = \{A_1, \ldots, A_M\}$ decides for i if and only if $X \in A_i$ (i.e., $s(X) = i$),

$$P(\hat{\Theta} \neq i | \Theta = i) = P(X \in \bar{A}_i | \Theta = i) = 1 - P(X \in A_i | \Theta = i).$$

Therefore,

$$P_E = 1 - \sum_{i=1}^{M} P(X \in A_i | \Theta = i) \pi(i). \tag{57}$$

But, from Eq. (55), $P(X \in A_i | \Theta = i) = \int_{A_i} f_i(x)\, dx$. Therefore,

$$P_E = 1 - \sum_{i=1}^{M} \pi(i) \int_{A_i} f_i(x)\, dx,$$

equivalently,

$$P_E = 1 - \int_{\mathbb{R}^n} \left[\sum_{i=1}^{M} 1_{A_i}(x) \pi(i) f_i(x) \right] dx \tag{58}$$

where $1_A(x) = 1$ if $x \in A$, $= 0$ if $x \notin A$.

Recall our goal of minimizing P_E as a function of the partition $\{A_1, \ldots, A_M\}$. From Eq. (58) this amounts to maximizing

$$\int_{\mathbb{R}^n} \left[\sum_{i=1}^{M} 1_{A_i}(x) \pi(i) f_i(x) \right] dx.$$

We shall see that we are fortunate enough to be able to find a partition that maximizes $\sum_{i=1}^{M} 1_{A_i}(x) \pi(i) f_i(x)$ for each $x \in \mathbb{R}^n$, and this partition will do.

Let us see how this partition $\{A_1^*, \ldots, A_M^*\}$ is constructed. First we show that

$$A_1^* = \{x \in \mathbb{R}^n | \pi(1) f_1(x) \geqslant \max(\pi(2) f_2(x), \ldots, \pi(M) f_M(x))\}. \tag{59}$$

More precisely, if a strategy s is such that $s^{-1}(1) = A_1$ and if A_1 is not identical to A_1^*, one can find another strategy $\bar{s}$ with $\bar{s}^{-1}(1) = A_1^*$ which is at least as good as s.

Indeed, suppose that there exists $u \in A_1$, $u \notin A_1^*$. Since $u \notin A_1^*$, there exists $i_0 \neq 1$ such that $\pi(i_0) f_{i_0}(u) > \pi(1) f_1(u)$. Clearly, the strategy $\bar{s}$ obtained from $s = \{A_1, \ldots, A_M\}$ by transporting u from A_1 to A_{i_0}, i.e., $\bar{s} = \{A_1 - \{u\}, A_2, \ldots, A_{i_0} + \{u\}, A_M\}$, is at least as good as s. Therefore, we can assume that for an optimal strategy s^*, A_1^* is given by Eq. (59).

Illustration 7. An Introduction to Bayesian Decision Theory   153

The same line of argument will show that we may assume that

$$A_2^* = \{x \in \mathbb{R}^n | x \notin A_1^* \quad \text{and} \quad \pi(2)f_2(x) \geq \max(\pi(3)f_3(x), \ldots, \pi(M)f_M(x))\}.$$

More generally, the A_i^*'s are defined recursively by

$$A_i^* = \{x \in \mathbb{R}^n | x \notin A_1^* \cup \cdots \cup A_{i-1}^* \quad \text{and}$$

$$\pi(i)f_i(x) \geq \max(\pi(i+1)f_{i+1}(x), \ldots, \pi(M)f_M(x))\}.$$

This amounts to

$$A_i^* = \{x \in \mathbb{R}^n | \pi(i)f_i(x) > \max_{j=1,\ldots,i-1} (\pi(j)f_j(x),$$

$$\pi(i)f_i(x)) \geq \max_{j=i+1,\ldots,M} (\pi(j)f_j(x))\}. \tag{60}$$

One can also say that the optimal decision region in favor of hypothesis i is the set of $x \in \mathbb{R}^n$ such that the quantity $\pi(i)f_i(x)$ is strictly larger than all quantities $\pi(j)f_j(x)$, for all $j \neq i$, $1 \leq j \leq M$. This is not quite what Eq. (60) says, but it is not a problem if the set C_i of $x \in \mathbb{R}^n$ such that $\pi(i)f_i(x) = \max_{j=i+1,\ldots,M} \pi(j)f_j(x)$ is such that $\int_{C_i} f_i(x) \, dx = 0$. Such is the case in the telemetry problem to which we now return.

Gaussian Hypotheses. In the telemetry problem $\pi(i) = $ constant, so that the optimal strategy which is

$$\text{When} \quad X \in A_i^* \quad \text{decide for} \quad i \tag{61a}$$

reads

$$\text{If} \quad f_i(X) > f_j(X) \quad \text{for all} \quad j \neq i, \quad \text{decide for} \quad i. \tag{61b}$$

Thus, for an observed random vector X, we have to find the index i that maximizes $f_i(X)$, or any nondecreasing function g of $f_i(X)$. Or equivalently, we have to find the index i that minimizes any nonincreasing function h of $f_i(X)$.

In view of the special shape of $f_i(x)$ given by Eq. (56), we see that the optimal decision rule reads

$$\text{If} \quad \|X - m_i\|^2 < \|X - m_j\|^2 \quad \text{for all} \quad j \neq i, \quad \text{decide for} \quad i. \tag{61c}$$

Since $\|X - m\|^2 = \|X\|^2 - 2m'X + \|m\|^2$, a further simplification is possible.

$$\text{If} \quad 2m_i'X - \|m_i\|^2 > 2m_j'X - \|m_j\|^2 \quad \text{for all} \quad j \neq i, \quad \text{decide for} \quad i. \tag{61d}$$

Recall that $m_i = (0, 0, \ldots, 0, \sigma_0, \sigma_1, \ldots, \sigma_L, 0, \ldots, 0)$ and therefore $\|m_i\|^2 = \sigma_0^2 + \cdots + \sigma_L^2$ does not depend on i. The decision rule is therefore

$$\text{If} \quad m_i'X > m_j'X \quad \text{for all} \quad j \neq i, \quad \text{decide for} \quad i. \tag{62}$$

A mathematician may be content with such a simple and elegant solution. But then we must think of the sonar operator, who either did not major in mathematics or has a deep aversion for computing scalar products of vectors of size exceeding 100. Here is what can be done to make this work easier.

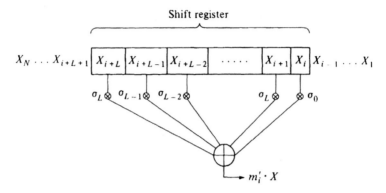

Figure 14

Consider the device pictured in Fig. 14. It consists of a "shift register" in which you stuff the vector X. The shift register contains $L + 1$ boxes, and at each click of a clock, the content of the register is shifted to the right. As time runs, the situation in the shift register evolves as follows (with $L = 4$):

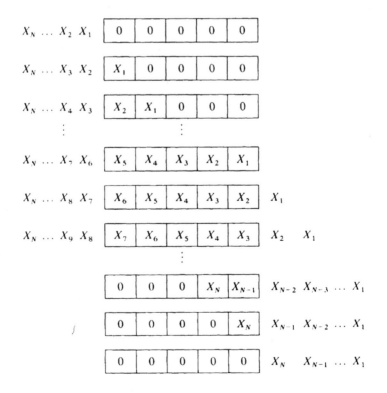

Illustration 7. An Introduction to Bayesian Decision Theory 155

Therefore, in Fig. 14, we have the situation at time $(i + L)\Delta$.

The content of the Kth from the right box of the register is multiplied by σ_{K-1}, and all the results are summed up to yield

$$m_i' X = \sigma_0 X_i + \sigma_1 X_{i+1} + \cdots + \sigma_L X_{i+L}. \tag{63}$$

The output of the device of Fig. 14 at time $(i + L)\Delta$ is therefore $m_i' X$. It can be presented to the sonar operator on a screen, on which the curve $i \to m_i' X$ will progressively shape itself. The maximum of this curve will directly give the distance to be measured if the "time" axis is correctly scaled in distance (Fig. 15).

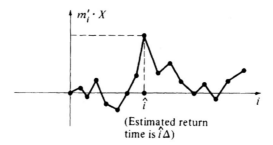

(Estimated return time is $\hat{i}\Delta$)

Figure 15. The processed waveform.

Recall that the σ_n's and X_n's are samples of $(\sigma(t), t \geqslant 0)$ and $(X(t), t \geqslant 0)$. In particular,

$$\sigma_0 X_i + \cdots + \sigma_L X_{i+L} = \sigma(0)X(i\Delta) + \cdots + \sigma(L\Delta)X((i + L)\Delta),$$

and we see that if Δ is very small, the right-hand side of the above equality is an approximation of

$$\int_0^a \sigma(t)X(i\Delta + t)\,dt = \int_{-\infty}^{+\infty} \sigma(t)X(i\Delta + t)\,dt.$$

(Recall that $a = L\Delta$ and that $\sigma(t)$ is null outside $[0, a]$.) Letting $\check{\sigma}(t) = \sigma(-t)$, we see that

$$\int_{-\infty}^{+\infty} \sigma(t)X(i\Delta + t)\,dt = \int_{-\infty}^{+\infty} \check{\sigma}(t - i\Delta)X(t)\,dt.$$

The purpose of obtaining this particular form is the following: there exist devices known as homogeneous linear filters of impulse response $(h(t), t \in \mathbb{R})$ that transform an "input" $(X(t), t \in \mathbb{R})$ into an "output" $(Y(t), t \in \mathbb{R})$ given by

$$Y(t) = \int_{-\infty}^{+\infty} h(t - s)X(s)\,ds.$$

We now see that what the sonar operator sees is the output of the homogeneous linear filter of impulse response $\sigma(-t)$ fed by the input $(X(t), t \in \mathbb{R})$ and sampled at time $i\Delta$. This particular filter is called the filter adapted to

$\sigma(t)$, and it is a notion of central importance in Detection and Communications theory.

In conclusion, we shall treat the more realistic case where the return signal is attenuated as a function of the distance. It is not difficult to see that it suffices to replace $m_i = (0, \ldots, 0, \sigma_0, \ldots, \sigma_L, 0, \ldots, 0)$ by $\tilde{m}_i = h(i)m_i$ where $h(i)$ is a scalar representing the attenuation factor. If attenuation is inversely proportional to distance, then $h(i) = \alpha/i$ for some constant α. The decision rule (62) must then be replaced by if $\tilde{m}'_i X - \frac{1}{2}\|\tilde{m}_i\|^2 > \tilde{m}^1_j X - \frac{1}{2}\|\tilde{m}_j\|^2$ for all $j \neq i$, decide i. But $\|\tilde{m}_i\|^2 = h(i)^2 \|m_i\|^2 = h(i)^2 E_\sigma$ where E_σ is the energy of the emitted bip, i.e., $\sigma_0^2 + \cdots + \sigma_L^2$. Therefore, the decision rule consists in comparing the quantities $m'_i X - \frac{1}{2}h(i)E_\sigma \qquad (1 \leqslant i \leqslant M)$.

SOLUTIONS FOR CHAPTER 4

E1. Recall that F is an increasing function right continuous with left-hand limits, such that $F(-\infty) = 0$, $F(+\infty) = 1$. We can therefore define the inverse function

$$F^{-1}(y) = \inf\{x \,|\, F(x) \geqslant y\} \qquad (y \in (0, 1)).$$

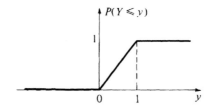

Definition of $F^{-1}(y)$ when F presents a step at y.

We have $F(F^{-1}(y)) = y$ for all $y \in (0, 1)$ and $F(x) \leqslant y \Leftrightarrow x \leqslant F^{-1}(y)$ for all x, y in the appropriate domains. Therefore, if

$$y \in (0, 1): P(Y \leqslant y) = P(F(X) \leqslant y) = P(X \leqslant F^{-1}(y))$$

$$= F(F^{-1}(y)) = y.$$

Of course, $P(Y \leqslant y) = 0$ if $y \leqslant 0$, and $= 1$ if $y \geqslant 1$.

E2. Since $Y \geq 0$, $F_Y(y) = P(Y \leq y) = 0$ when $y \leq 0$. If $y \geq 0$,

$$F_Y(y) = P(X^2 \leq y) = P(-\sqrt{y} \leq X \leq +\sqrt{y})$$

$$= \frac{1}{\sqrt{2\pi}} \int_{-\sqrt{y}}^{+\sqrt{y}} e^{-z^2/2} \, dz = \sqrt{\frac{2}{\pi}} \int_0^{\sqrt{y}} e^{-z^2/2} \, dz.$$

Therefore, for $y > 0$,

$$f(y) = \frac{dF}{dy}(y) = \sqrt{\frac{2}{\pi}} e^{-y/2} \times \frac{1}{2} y^{-1/2}.$$

E3. From E2, we see that $X_i^2 \sim \gamma(1/2, 1/2)$. From Exercise E20 of Chapter 3, we get $\sum_{i=1}^n X_i^2 \sim \gamma(n/2, 1/2) = \chi_n^2$.

E4.
$$P(Z \leq x) = P(UY \leq x) = P(U = 1, Y \leq x) + P(U = -1, Y \geq -x)$$

$$= P(U = 1)P(Y \leq x) + P(U = -1)P(Y \geq -x)$$

$$= \tfrac{1}{2}P(Y \leq x) + \tfrac{1}{2}P(Y \geq -x)$$

$$= P(Y \leq x) \quad \text{since} \quad P(Y \leq x) = P(Y \geq -x).$$

$E[YZ] = E[UY^2] = E[U]E[Y^2] = 0$, i.e. Y and Z are uncorrelated. If Y and Z are independent, the vector (Y, Z) admits a probability density and thus $Y + Z$ admits a probability density. Therefore, $P(Y + Z = 0)$ should be null, a contradiction of the fact that

$$P(Y + Z = 0) = P(Y \cdot (1 + U) = 0) = P(\{Y = 0\} \cup \{1 + U = 0\})$$

$$= P(1 + U = 0) = \tfrac{1}{2}.$$

E5. $\sigma_{12} = \sigma_{21} = E[X_1 X_2]$. The nondegeneracy condition is $\det \Gamma = \sigma_1^2 \sigma_2^2 - \sigma_{12}\sigma_{21} > 0$, that is, $|\rho| > 0$ since $\rho^2 = \sigma_{12}^2/(\sigma_1^2 \sigma_2^2)$. The inverse of Γ is

$$\Gamma^{-1} = (\sigma_1^2 \sigma_2^2 - \sigma_{21}^2)^{-1} \begin{pmatrix} \sigma_2^2 & -\sigma_{21} \\ -\sigma_{12} & \sigma_1^2 \end{pmatrix},$$

and therefore,

$$x'\Gamma^{-1}x = \frac{\dfrac{x_1^2}{\sigma_1^2} - 2\rho \dfrac{x_1 x_2}{\sigma_1 \sigma_2} + \dfrac{x_2^2}{\sigma_2^2}}{1 - \rho^2} = Q(x_1, x_2).$$

Let

$$\begin{cases} z_1 = \left(\dfrac{x_1}{\sigma_1} - \rho \dfrac{x_2}{\sigma_2}\right) \dfrac{1}{\sqrt{1 - \rho^2}} \\ \\ z_2 = \dfrac{x_2}{\sigma_2}. \end{cases}$$

Then

$$Q(x_1, x_2) = \frac{\left(\dfrac{x_1}{\sigma_1} - \rho \dfrac{x_2}{\sigma_2}\right)^2}{1 - \rho^2} + \left(\dfrac{x_2}{\sigma_2}\right)^2 = z_1^2 + z_2^2.$$

Therefore, if we define

$$\begin{cases} Z_1 = \left(\dfrac{X_1}{\sigma_1} - \rho\dfrac{X_2}{\sigma_2}\right)\dfrac{1}{\sqrt{1-\rho^2}} \\[2mm] Z_2 = \dfrac{X_2}{\sigma_2}, \end{cases}$$

then (Z_1, Z_2) is a standard Gaussian vector.

E.6. With the notations of Eqs. (66), (67), and (68) of Chapter 3, Section 2.2, $X = DY$ where

$$D = C\binom{I_r}{0}\begin{pmatrix} \sigma_1 & & & & \bigcirc \\ & \ddots & & & \\ & & \sigma_r & & \\ & & & 0 & \\ & \bigcirc & & & \ddots \\ & & & & 0 \end{pmatrix},$$

and I_r is the $(r \times r)$ identity matrix.

E.7. $$E(X_1^2 X_2^2) = E[X_1 X_1 X_2 X_2] = \sigma_{11}\sigma_{22} + \sigma_{12}\sigma_{12} + \sigma_{12}\sigma_{12}$$

$$= \sigma_1^2\sigma_2^2 + 2\sigma_{12}^2 = 3$$

$$E[X^4] = 3\sigma^4 = 3.$$

E.8. $$f_{X,Y}(x,y) = f_X(x)f_Y(y)$$

$$= \frac{1}{2^{(m+n)/2}\Gamma\left(\dfrac{n}{2}\right)\Gamma\left(\dfrac{m}{2}\right)} x^{(n/2)-1} y^{(m/2)-1} e^{-(x+y)/2} \qquad (x \geqslant 0, y \geqslant 0)$$

$$(Z, Y) = g(X, Y)$$

where

$$g = U \subset \mathbb{R}^2 \to \mathbb{R}^2, \qquad U = \{x > 0, y > 0\}, \qquad P((X, Y) \in U) = 1.$$

We have

$$V = g(U) = \{z > 0, y > 0\}, \qquad g^{-1}(z, y) = (zy, z), \qquad \left|\frac{Dg^{-1}(z,y)}{D(z,y)}\right| = |y|.$$

Therefore,

$$f_{Z,Y}(z,y) = \frac{1}{2^{(m+n)/2}\Gamma\left(\dfrac{n}{2}\right)\Gamma\left(\dfrac{m}{2}\right)} z^{(n/2)-1} y^{[(m+n)/2]-1} e^{-y[(1+z)/2]} \qquad (z > 0, y > 0).$$

The density of Z is obtained by integration of $f_{Z,Y}(z,y)$ with respect to y. Since

$$\int_0^\infty y^{[(m+n)/2]-1} e^{-y[(1+z)/2]} dy$$

$$= \frac{2^{(m+n)/2}}{(1+z)^{(m+n)/2}} \int_0^\infty u^{[(m+n)/2]-1} e^{-u} du = \frac{2^{(n+m)/2}\Gamma\left(\dfrac{n+m}{2}\right)}{(1+z)^{(n+m)/2}},$$

$$f_Z(z) = \frac{\Gamma\left(\dfrac{m+n}{2}\right)}{\Gamma\left(\dfrac{n}{2}\right)\Gamma\left(\dfrac{m}{2}\right)} \frac{z^{(n/2)-1}}{(1+z)^{(m+n)/2}} \qquad (z > 0).$$

E9. With the notation of Exercise E8 $(m/n)Z$ admits the density $(n/m)f_Z((n/m)x)$, which is simply the Fisher density with parameters m and n. The result follows from this and Theorem T3.

E10. With the notations of Exercise E8,

$$f_{\sqrt{Z}}(x) = f_Z(x^2)2x \qquad (x \geqslant 0)$$

(change of variable $x = \sqrt{z}$). Therefore,

$$f_{\sqrt{Z}}(x) = \frac{2\Gamma\left(\dfrac{m+n}{2}\right)}{\Gamma\left(\dfrac{m}{2}\right)\Gamma\left(\dfrac{n}{2}\right)} \frac{x^{n-1}}{(1+x^2)^{(n+m)/2}} \qquad (x > 0),$$

and this is indeed a Fisher–Snedecor probability density.

E11. $(1/\sqrt{n})(\bar{X} - m) \sim \mathcal{N}(0, 1)$ and therefore its square is χ_1^2, $(n-1)|S|^2 \sim \chi_{n-1}^2$. Therefore $|T| \sim \sqrt{n-1}\sqrt{(\chi_1^2/\chi_{n-1}^2)}$. By E10, $|T|\sqrt{n-1}$ is therefore a Fisher–Snedecor r.v. with parameters 1 and $n-1$. Since T has a symmetric distribution $f_T(t) = \frac{1}{2}f_{|T|}(|t|)$, and this is the Student distribution with parameter $n-1$.

E12. From Eq. (32) of Chapter 3, if $X \sim \mathscr{E}(\lambda)$,

$$\phi_X(u) = \frac{\lambda}{\lambda - iu}.$$

Therefore, since the X_i's are independent,

$$\phi_{X_1 + \cdots + X_n}(u) = \prod_{i=1}^{n} \phi_{X_i}(u) = \left(\frac{\lambda}{\lambda - iu}\right)^n.$$

Now, from Eq. (34) of Chapter 3, $[\lambda/(\lambda - iu)]^n$ is the characteristic function of a random variable with a gamma density with parameter $\alpha = n$, $\beta = \lambda$.

E13. Let $t_1, \ldots, t_k, h_1, \ldots, h_k$ be such that

$$0 < t_1 < t_1 + h_1 < t_2 < t_2 + h_2 < \cdots < t_{k-1} + h_{k-1} < t_k < t_k + h_k.$$

Denote

$$A = \{T_i \in [t_i + h_i], i = 1, \ldots, k\}.$$

Clearly

$$A = \{N(t_1) = 0, N(t_1 + h_1) - N(t_1) = 1, \ldots, N(t_{k-1} + h_{k-1}) - N(t_{k-1}) = 1,$$
$$N(t_k) - N(t_{k-1} + h_{k-1}) = 0, N(t_k + h_k) - N(t_k) \geqslant 1\}.$$

Therefore,

$$P(A) = P(N(t_1) = 0)P(N(t_1 + h_1) - N(t_1) = 1)\dots$$

$$P(N(t_{k-1} + h_{k-1}) - N(t_k) = 0)P(N(t_k + h_k) - N(t_k) \geq 1)$$

$$= e^{-\lambda t_1}e^{-\lambda h_1}\lambda h_1 \dots e^{-\lambda(t_k - t_{k-1} - h_{k-1})}(1 - e^{-\lambda h_k})$$

$$= \lambda^k e^{-\lambda t_k} h_1 h_2 \dots h_{k-1} h_k \frac{1 - e^{-\lambda h_k}}{h_k}.$$

But $\lim_{h_k \downarrow 0}(1 - e^{-\lambda h_k})/h_k = 1$, therefore

$$\lim_{h_1, \dots, h_k \downarrow 0} \frac{P(T_i \in [t_i, t_i + h_i], i = 1, \dots, k)}{h_1 h_2 \dots h_k} = \lambda^k e^{-\lambda t_k}.$$

This is true under the condition $0 < t_1 < t_2 < \cdots < t_k$. Otherwise the limit is 0 since $0 < T_1 < \cdots < T_k$, P-as (give the complete argument for this). Finally,

$$f_{T_1, \dots, T_k}(t_1, \dots, t_k) = \lambda^k e^{-\lambda t_k} 1_D(t_1, \dots, t_k)$$

where $D = \{(t_1, \dots, t_k) \in R^k | 0 < t_1 < \cdots < t_k\}$.

Now

$$S_1 = T_1$$

$$S_2 = T_2 - T_1$$

$$\vdots$$

$$S_k = T_k - T_{k-1}$$

i.e., $(S_1, \dots, S_k) = g(T_1, \dots, T_k)$. Observing that $g(D) = \{(s_1, \dots, s_k) \in R^k | 0 < s_1, \dots, 0 < s_k\}$ and that $|J_g(t)| = |J_{g^{-1}}(s)| = 1$, the formula of smooth change of variables yields:

$$f_{S_1, \dots, S_k}(s_1, \dots, s_k) = \prod_{i=1}^{k} (\lambda e^{-\lambda s_i}), \qquad s_i > 0 \quad (1 \leq i \leq k).$$

E14. Define

$$A = \{\omega | \exists k(\omega), l(\omega) \in N_+ \qquad \text{such that} \quad k(\omega) \neq l(\omega) \text{ and } T_k(\omega) = T_l(\omega)\}$$

$$A_n = \{\omega | \exists k(\omega), l(\omega) \in N_+ \qquad \text{such that} \quad k(\omega) \neq l(\omega) \text{ and } T_k(\omega) = T_l(\omega) \leq n\}.$$

Clearly $A_n \subset A$ and $A_n \uparrow A$, so that $P(A) = \lim_{n \uparrow \infty} P(A_n)$. It therefore suffices to prove $P(A_n) = 0$ for all $n \in N_+$. We will do the proof for $n = 1$ to simplify the notations.

Consider the event

$$B_k = \left[N\left(\frac{i+1}{2^k}\right) - N\left(\frac{i}{2^k}\right) \leq 1, i = 0, \dots, 2^k - 1 \right] \qquad (k \geq 1).$$

Clearly $B_{k+1} \supset B_k$ $(k \geq 1)$ and $A_1 \subset \bar{B}_k$ $(k \geq 1)$. Therefore, $P(A_1) \leq P(\bar{B}_k)$ $(k \geq 1)$ and $P(A_1) \leq \lim_{k \uparrow \infty} P(\bar{B}_k)$. Now

$$P(B_k) = \prod_{i=0}^{2^k - 1} P\left[N\left(\frac{i+1}{2^k}\right) - N\left(\frac{i}{2^k}\right) \leq 1 \right] = e^{-\lambda}\left(1 + \frac{\lambda}{2^k}\right)^{2^k}$$

and therefore $\lim_{k\uparrow\infty} P(B_k) = e^{-\lambda}e^{\lambda} = 1$. Finally, $0 \le P(A_1) \le \lim_{k\uparrow\infty} P(\bar{B}_k) = 0$ so that $P(A_1) = 0$.

E15.
$$F_{T_n}(t) = P(T_n \le t) = P[N(t) \ge n] = 1 - P[N(t) \le n-1]$$

$$= 1 - \sum_{k=0}^{n-1} \exp\left\{-\int_0^t \lambda(s)\,ds\right\} \frac{\left(\int_0^t \lambda(s)\,ds\right)^k}{k!}.$$

So that, for $t \ge 0, n \ge 1$,

$$f_{T_n}(t) = \frac{dF_{T_n}(t)}{dt} = \lambda(t)\exp\left\{-\int_0^t \lambda(s)\,ds\right\} \frac{\left(\int_0^t \lambda(s)\,ds\right)^{n-1}}{(n-1)!}.$$

In particular,

$$F_{T_1}(t) = 1 - \exp\left\{-\int_0^t \lambda(s)\,ds\right\} \quad \text{and} \quad f_{T_1}(t) = \lambda(t)\exp\left\{-\int_0^t \lambda(s)\,ds\right\}.$$

If $T_1 < \infty$ P-as, then $F_{T_1}(+\infty) = 1$, i.e. $\int_0^\infty \lambda(s)\,ds = \infty$.

E16.
$$E\left[\sum_{n=1}^\infty g(T_n, X_n)1_{\{T_n \le t\}}\right] = \sum_{n=1}^\infty E[g(T_n, X_n)1_{\{T_n \le t\}}].$$

Let f_{T_n} and f_{X_n} be the probability densities of T_n and X_n. Since T_n and X_n are independent, the probability density of (T_n, X_n) is $f_{T_n,X_n}(t,x) = f_{T_n}(t)f_{X_n}(x)$. Therefore,

$$E[g(T_n, X_n)1_{\{T_n \le t\}}] = \int_0^t \int_0^\infty g(s,x)f_{T_n}(s)f_{X_n}(x)\,ds\,dx$$

$$= \int_0^t E[g(s, X_1)]f_{T_n}(s)\,ds$$

since

$$E[g(s, X_1)] = \int_0^\infty g(s,x)f_{X_1}(x)\,dx = \int_0^\infty g(s,x)f_{X_n}(x)\,dx.$$

Therefore,

$$E\left[\sum_{n=1}^\infty g(T_n, X_n)1_{\{T_n \le t\}}\right] = \int_0^t E[g(s, X_1)]\left(\sum_{n=1}^\infty f_{T_n}(s)\right)ds.$$

But from E15, $\sum_{n\ge 1} f_{T_n}(t) = \lambda(t)$.

E17. We have

$$X(t) = \sum_{n\ge 1} 1_{\{T_n + X_n \ge t\}} 1_{\{T_n \le t\}}$$

Therefore, applying the result of Exercise E16 with $g(s,x) = 1_{\{s+x \ge t\}}$,

$$E[X(t)] = \int_0^t P(X \ge t - s)\lambda\,ds.$$

In the case where $X_1 \sim \mathscr{E}(\mu)$,

$$E[X(t)] = \frac{\lambda}{\mu}(1 - e^{-\mu t}).$$

In the general case

$$E[X(t)] = \lambda \int_0^t [1 - F(t - s)]\, ds = \lambda \int_0^t [1 - F(u)]\, du.$$

But

$$\int_0^t [1 - F(u)]\, du \xrightarrow[t \to \infty]{} E[X_1].$$

E18. The solution follows directly from the definitions by observing that the area of A_t is $\int_0^t \lambda(s)\, ds$.

E19. Observe that if $i = 1, \ldots, n$

$$\frac{p_i^{k_i}}{k_i!} = \frac{[\lambda S(A_i)]^{k_i}}{k_i!} \frac{1}{(D^2 \lambda)^{k_i}} = \frac{[\lambda S(A_i)]^{k_i}}{k_i!} \frac{1}{K^{k_i}}$$

and that

$$p_0^{k_0} = \left[1 - \sum_{i=1}^{n} \frac{S(A_i)}{D^2}\right]^{K - \sum_{i=1}^{n} k_i}$$

$$= \left\{1 - \left[\lambda \sum_{i=1}^{n} S(A_i)\right] \middle/ K\right\}^K \left[1 - \sum_{i=1}^{n} \frac{S(A_i)}{D^2}\right]^{-\sum_{i=1}^{n} k_i}$$

and therefore

$$\lim_{K \uparrow \infty} p_0^{k_0} = e^{-\lambda \sum_{i=1}^{n} S(A_i)}.$$

It therefore remains to prove that

$$\lim_{K \uparrow \infty} \prod_{i=1}^{n} \frac{1}{K^{k_i}} \frac{K!}{k_0!} = 1$$

But since $k_0 = K - \sum_{i=1}^{n} k_i$,

$$\frac{K!}{(K^{\sum k_i}) k_0!} = \frac{K(K-1)\ldots(K - \sum k_i + 1)}{KK\ldots K}$$

where the numerator contains $\sum k_i$ terms. The latter quantity obviously converges to 1 as $K \uparrow \infty$.

E20. Define $X_1 = W(t_1)$, $X_2 = W(t_2) - W(t_1)$, $\ldots$, $X_n = W(t_n) - W(t_{n-1})$ and $Y_1 = W(t_1)$, $Y_2 = W(t_2)$, $\ldots$, $Y_n = W(t_n)$. Therefore,

$$Y_1 = X_1, \qquad Y_2 = X_1 + X_2, \qquad Y_3 = X_1 + X_2 + X_3, \ldots, Y_n = X_1 + \cdots + X_n.$$

The Jacobian determinant corresponding to the function $g: (x_1, \ldots, x_n) \to (x_1, x_1 + x_2, \ldots, x_1 + \cdots + x_n)$ is 1. Since $g(\mathbb{R}^n) = \mathbb{R}^n$, the densities f_X and f_Y of $(X_1, \ldots, X_n)$ and $(Y_1, \ldots, Y_n)$ are related by $f_Y(y) = f_X[g^{-1}(y)]$, i.e.,

$$f_Y(y_1, \ldots, y_n) = f_X(y_1, y_2 - y_1, \ldots, y_n - y_{n-1}).$$

CHAPTER 5

Convergences

1. Almost-Sure Convergence

1.1. The Borel–Cantelli Lemma

The notion of almost-sure convergence of a sequence of random variables $(X_n, n \geq 1)$ toward a random variable X is a very natural one.

D1 Definition. One says that $X_n \xrightarrow{\text{as}} X$ (read "X_n converges to X almost surely when n goes to infinity") if there exists an event N of null probability such that for all ω outside N, $\lim_{n \uparrow \infty} X_n(\omega) = X(\omega)$, where the latter limit is in the ordinary sense.

Remark. If the almost-sure limit of a sequence $(X_n, n \geq 1)$ exists, it is *essentially unique*, that is, if $X_n \xrightarrow{\text{as}} X$ and $X_n \xrightarrow{\text{as}} X'$, then $X = X'$, P-as. Indeed, let N and N' be the events of null probability of Definition D1 corresponding to X and X', respectively. Outside $N \cup N'$, $\lim_{n \uparrow \infty} X_n(\omega) = X(\omega)$ and $\lim_{n \uparrow \infty} X_n(\omega) = X'(\omega)$, therefore, if $\omega \notin N \cup N'$, $X(\omega) = X'(\omega)$. Since $P(N \cup N') \leq P(N) + P(N') = 0$, we see that $X = X'$, P-as.

Now suppose that $X_n \xrightarrow{\text{as}} X$ and select an arbitrarily small number, say 10^{-6}. Consider the events $A_n = \{|X_n - X| \geq 10^{-6}\}$. In view of the definition of almost-sure convergence, if $\omega \notin N$, $|X_n(\omega) - X(\omega)|$ can be larger than 10^{-6} only for a finite (perhaps empty) set of indices n. In other words,

$$P(\omega \in A_n \quad \text{infinitely often}) = 0.$$

This remark provides motivation for the study of the set $\{A_n \quad \text{i.o.}\} = \{\omega | \omega \in$

A_n infinitely often} which we now define for an arbitrary sequence of events $(A_n, n \geq 1)$ in terms of the usual elementary set operations as follows:

$$\{A_n \quad \text{i.o.}\} = \bigcap_{n \geq 1} \bigcup_{k \geq n} A_k. \tag{1}$$

The next exercise shows that this is the relevant definition.

E1 Exercise. Prove that $\omega \in \{A_n \quad \text{i.o.}\}$ if and only if there are infinitely many integers n such that $\omega \in A_n$.

The so-called Borel–Cantelli lemma is the key to many problems concerning almost-sure convergence, as we shall soon see. It consists of two parts.

T2 Theorem (Borel–Cantelli, Direct Part). *For any sequence of events* $(A_n, n \geq 1)$,

$$\sum_{n \geq 1} P(A_n) < \infty \Rightarrow P(A_n \quad \text{i.o.}) = 0. \tag{2}$$

PROOF OF EQ. (2). Define $B_n = \bigcup_{k \geq n} A_k$. Clearly $(B_n, n \geq 1)$ is decreasing, and therefore, by the sequential continuity of P and Definition D1, $P(A_n \quad \text{i.o.}) = \lim_{n \uparrow \infty} \downarrow P(\bigcup_{k \geq n} A_k)$. But by the sub-$\sigma$-additivity of P, $P(\bigcup_{k \geq n} A_k) \leq \sum_{k \geq n} P(A_k)$. The conclusion follows from the hypothesis $\sum_{n \geq 1} P(A_n) < \infty$. $\quad\square$

The converse part of Borel–Cantelli requires *independence* of the family $(A_n, n \geq 1)$.

T3 Theorem (Borel–Cantelli, Converse Part). *If the events* $A_n(n \geq 1)$ *are independent*

$$\sum_{n \geq 1} P(A_n) = \infty \Rightarrow P(A_n \quad \text{i.o.}) = 1. \tag{3}$$

PROOF OF EQ. (3). Since the A_n's are independent, for all $n \geq 1$

$$P\left(\bigcap_{k \geq n} \bar{A}_k\right) = \prod_{k \geq n} P(\bar{A}_k) = \prod_{k \geq n} [1 - P(A_k)].$$

We can suppose, without loss of generality, that $0 < P(A_n) < 1$ for all $n \geq 1$ (why?). By a standard result concerning infinite products, $\prod_{k \geq n} [1 - P(A_k)] = 0$ is equivalent to $\sum_{k \geq n} P(A_k) = \infty$. Therefore, for all $n \geq 1$, $P(\bigcap_{k \geq n} \bar{A}_k) = 0$, or equivalently (by de Morgan's rule) $P(\bigcup_{k \geq n} A_k) = 1$. But by the sequential continuity of probability $P(A_n \quad \text{i.o.}) = P(\bigcap_{n \geq 1} \bigcup_{k \geq n} A_k) = \lim_{n \uparrow \infty} \downarrow P(\bigcup_{k \geq n} A_k) = 1$. $\quad\square$

EXAMPLE 1. Let $(X_n, n \geq 1)$ and $(Y_n, n \geq 1)$ be two independent sequences of random variables, each being iid. Assume that $P(X_1 \geq Y_1) > 0$. The events

$A_n = \{X_n \geq Y_n\}$ are independent with $P(A_n) = P(X_n \geq Y_n) > 0$ and therefore $\sum_{n \geq 1} P(X_n \geq Y_n) = \infty$. Thus $P(X_n \geq Y_n \quad \text{i.o.}) = 1$.

E2 Exercise. Let $(X_n, n \geq 1)$ be a sequence of independent random variables taking the values 0 or 1. Let $p_n = P(X_n = 1)$. Prove that $X_n \xrightarrow{\text{as}} 0$ if and only if $\sum_{n \geq 1} p_n < \infty$.

1.2. A Criterion for Almost-Sure Convergence

T4 Theorem (Criterion of Almost-Sure Convergence). *Let $(X_n, n \geq 1)$ be a sequence of random variables. It converges almost surely to the random variable X if and only if*

$$P(|X_n - X| \geq \varepsilon \quad \text{i.o.}) = 0 \qquad \text{for all} \quad \varepsilon > 0. \tag{4}$$

PROOF OF EQ. (4). The direct part was proven in Section 1.1. For the converse part, assume that Eq. (4) holds and define for $k \geq 1$ the event $N_k = \{|X_n - X| \geq 1/k \quad \text{i.o.}\}$. By assumption $P(N_k) = 0$ so that $P(N) = 0$, where $N = \bigcup_{k \geq 1} N_k$. If $\omega \notin N$, then $\omega \notin N_k$ for all $k \geq 1$, and therefore, by definition of N_k, the set of indices n such that $|X_n(\omega) - X(\omega)| \geq 1/k$ is empty or finite for all $k \geq 1$. This implies that if $\omega \notin N$, $\lim_{n \uparrow \infty} X_n(\omega) = X(\omega)$. $\qquad \square$

In view of proving almost-sure convergence, criterion (4) is generally used in conjunction with the direct part of the Borel–Cantelli lemma, and some upper bound of the Markov–Chebyshev type. The idea is to prove that

$$\sum_{n \geq 1} P(|X_n - X| \geq \varepsilon) < \infty \qquad \text{for all} \quad \varepsilon > 0 \tag{5}$$

which implies Eq. (4) (by Borel–Cantelli). The proof of Eq. (5) generally consists in upper bounding $P(|X_n - X| \geq \varepsilon)$ by the general term of a convergent series.

EXAMPLE 2 (Borel's Law of Large Numbers). Consider a sequence of independent random variables $(X_n, n \geq 1)$ with values in $\{0, 1\}$ such that

$$P(X_n = 1) = p \qquad (n \geq 1).$$

Define the empirical frequency of "1."

$$\bar{X}_n = \frac{X_1 + \cdots + X_n}{n} = \frac{S_n}{n}.$$

As expected, $\bar{X}_n \xrightarrow{\text{as}} p$. To prove this result due to Borel (1909), the above program will be applied. We therefore seek to prove that

$$\sum_{n \geq 1} P\left[\left|\frac{S_n}{n} - p\right| \geq \varepsilon\right] < \infty.$$

E3 Exercise. Apply Markov's inequality to obtain the announced result.

E4 Exercise. Let $(X_n, n \geq 1)$ be a sequence of random variables, and let X be another random variable. Show that if $P(|X_n - X| \geq \varepsilon_n) \leq \delta_n$ for some non-negative sequences $(\varepsilon_n, n \geq 1)$ and $(\delta_n, n \geq 1)$ such that $\lim_{n \uparrow \infty} \varepsilon_n = 0$ and $\sum_{n \geq 1} \delta_n < \infty$, then $X_n \xrightarrow{\text{as}} X$.

A general form of the strong law of large numbers will now be proven by essentially the same technique as the one used to prove Borel's result (Example 2).

1.3. The Strong Law of Large Numbers

T5 Theorem. *Let $(X_n, n \geq 1)$ be a sequence of identically distributed random variables. Assume that their mean $\mu = E[X_1]$ is defined and that they have a finite variance σ^2. Assume, moreover, that they are uncorrelated, i.e.,*

$$E[(X_i - \mu)(X_j - \mu)] = 0 \qquad \text{if} \quad i \neq j. \tag{6}$$

Then, letting $S_n = X_1 + \cdots + X_n$,

$$\frac{S_n}{n} \xrightarrow{\text{as}} \mu. \tag{7}$$

PROOF OF EQ. (7). There is no loss of generality in supposing that $\mu = 0$.

Denote $Z_m = \sup_{1 \leq k \leq 2m+1}(|X_{m^2+1} + \cdots + X_{m^2+k}|)$. Defining for each $n > 1$, the integer $m(n)$ by

$$m(n)^2 < n \leq [m(n) + 1]^2,$$

we have

$$\left| \frac{S_n}{n} \right| \leq \left| \frac{S_{m(n)^2}}{m(n)^2} \right| + \frac{Z_{m(n)}}{m(n)^2}.$$

Since $\lim_{n \uparrow \infty} m(n) = +\infty$, it suffices to prove that

$$\frac{S_{m^2}}{m^2} \xrightarrow{\text{as}} 0 \qquad \text{as} \quad m \to \infty \tag{$*$}$$

and

$$\frac{Z_m}{m^2} \xrightarrow{\text{as}} 0 \qquad \text{as} \quad m \to \infty. \tag{$**$}$$

For any $\varepsilon > 0$, by Chebyshev's inequality,

$$P\left(\left| \frac{S_{m^2}}{m^2} \right| \geq \varepsilon \right) \leq \frac{\text{Var}(S_{m^2})}{m^4 \varepsilon^2} = \frac{m^2 \sigma^2}{m^4 \varepsilon^2} = \frac{1}{m^2} \frac{\sigma^2}{\varepsilon^2}$$

(here we have used the fact that the X_i's are not correlated, which implies $\text{Var}[\sum_{i=1}^{k}(X_i)] = \sum_{i=1}^{k}\text{Var}(X_i)]$.) Therefore, $\sum_{m\geq 1} P(|S_{m^2}/m^2| \geq \varepsilon) < \infty$, and the conclusion $(*)$ follows by Borel–Cantelli's lemma and the criterion of almost-sure convergence, as explained in Section 1.2.

Denote $\xi_k = X_{m^2+1} + \cdots + X_{m^2+k}$. Clearly, if $|Z_m| \geq m^2\varepsilon$, then for at least one k, $1 \leq k \leq 2m+1$, $|\xi_k| \geq m^2\varepsilon$. Therefore,

$$\left\{\frac{Z_m}{m^2} \geq \varepsilon\right\} \subset \bigcup_{k=1}^{2m+1} \{|\xi_k| \geq m^2\varepsilon\}$$

and

$$P\left(\frac{Z_m}{m^2} \geq \varepsilon\right) \leq P\left(\bigcup_{k=1}^{2m+1} |\xi_k| \geq m^2\varepsilon\right) \leq \sum_{k=1}^{2m+1} P(|\xi_k| \geq m^2\varepsilon)$$

Using Chebyshev's inequality

$$P\left(\frac{Z_m}{m^2} \geq \varepsilon\right) \leq \sum_{k=1}^{2m+1} \frac{\text{Var}(\xi_k)}{m^4\varepsilon^2}.$$

Now $\text{Var}(\xi_k) = \sum_{i=1}^{k}\text{Var}(X_{m^2+i}) \leq (2m+1)\sigma^2$ when $k \leq 2m+1$ and therefore,

$$P\left(\frac{Z_m}{m^2} \geq \varepsilon\right) \leq \frac{(2m+1)^2}{m^4}\frac{\sigma^2}{\varepsilon^2}.$$

This inequality implies $\sum_{m\geq 1} P(Z_m/m^2 \geq \varepsilon) \leq \infty$ and Eq. $(**)$ follows by Borel–Cantelli's lemma and the criterion of almost-sure convergence, as above. $\qquad\square$

E5 Exercise. Let $(N(t), t \in \mathbb{R}_+)$ be the counting process associated with a homogeneous Poisson process $(T_n, n \geq 1)$ over the positive half-line. Prove that $\lim_{t\uparrow\infty} N(t)/t = \lambda$ a.s., where λ is the intensity (>0) of the point process.

E6 Exercise. A "pattern" is defined as a finite sequence of 0's and 1's, say $(x_0,\ldots,x_k)$. Let now $(X_n, n \geq 1)$ be an iid sequence of random variables with $P(X_n = 0) = P(X_n = 1) = \frac{1}{2}$, that is, a Bernoulli sequence. Define $Y_n = 1$ if $X_n = x_0$, $X_{n+1} = x_1$, $\ldots$, $X_{n+k} = x_k$ and $Y_n = 0$ otherwise. Show that $(Y_1 + \cdots + Y_n)/n \xrightarrow{\text{as}} (\frac{1}{2})^{k+1}$.

Remark. The term $(Y_1 + \cdots + Y_n)/n$ is the empirical frequency of "pattern coincidences." Therefore, in a Bernoulli sequence, which is the archetype of a "completely random" sequence of 0's and 1's, the "pattern coincidence law" holds. However, there are some deterministic sequences of 0's and 1's that are highly structured and "do not look random" at all and still verify the pattern coincidence law. This is the case, for instance, for the Champernowne sequence

01101110010111011110001001101010101111001101...,

which consists of a concatenation of the binary sequences representing 0, 1, 2, 3, 4, 5, 6, 7, ... in base 2. (The proof is not immediate!)

Remark. The form of the strong law of large numbers that is usually given in textbooks is the one due to Kolmogorov: Let $(X_n, n \geq 1)$ be a sequence of *independent* and identically distributed random variables with mean μ. Then Eq. (7) holds.

The differences with the strong law of large numbers proven above are

(i) in Kolmogorov's law, there is no assumption of finite variance;
(ii) however, the independence assumption is stronger than the noncorrelation assumption.

For practical purposes, the law of large numbers proven above is more interesting because noncorrelation is a property easier to test than independence. However, Kolmogorov's version is more useful in theoretical situations where the assumption of finite variance is not granted. Its proof will not be given here, but this should be of minor concern to us since whatever version of the strong law of large numbers is available, the loop is now closed. The paradigm of Probability Theory has been found consistent with the intuitive view of probability as an idealization of empirical frequency.

Although the principle of probabilistic modeling can now be accepted without discussion about its internal coherence, not all has been done. In practice, each model must be tested against reality and, if necessary, altered for a better fit with the data. This is the domain of Statistics.

Since probability is just asymptotic frequency, clearly the statistician will feel comfortable with large samples, that is, data originating from many repeated independent experiments. However large the sample, though, a statistician cannot expect the empirical frequency to be precisely equal to the probability. Thus, one needs to discriminate "acceptable" and "unacceptable" discrepancies between the (measured) empirical frequency and the (guessed) probability. The essential tool for doing this is the theory of convergence in law.

2. Convergence in Law

2.1. Criterion of the Characteristic Function

D6 Definition. Consider a sequence $(F_n, n \geq 1)$ of cumulative distribution functions on $\mathbb{R}$ and let F be another cdf on $\mathbb{R}$. One says that $(F_n, n \geq 1)$ converges weakly to F if

$$\lim_{n \uparrow \infty} F_n(x) = F(x) \qquad \text{whenever} \quad F(x) = F(x-). \tag{8}$$

In other words, $(F_n, n \geqslant 1)$ converges simply to F at the continuity points of F. This is denoted by

$$F_n \overset{w}{\rightarrow} F.$$

D7 Definition. Let $(X_n, n \geqslant 1)$ and X be real random variables with respective cdf $(F_{X_n}, n \geqslant 1)$ and F_X. One says that $(X_n, n \geqslant 1)$ converges in law (or in distribution) to X if $F_{X_n} \overset{w}{\rightarrow} F_X$, that is,

$$\lim_{n \uparrow \infty} P(X_n \leqslant x) = P(X \leqslant x) \qquad \text{whenever} \quad P(X = x) = 0. \tag{9}$$

This is denoted by

$$X_n \overset{\mathscr{L}}{\longrightarrow} X \quad \text{or} \quad \mathscr{L}(X_n) \to \mathscr{L}(X)$$

where $\mathscr{L}(Y)$ means the "law of Y", i.e., the "distribution of Y".

D8 Definition. Let $(X_n, n \geqslant 1)$ be a sequence of real random variables and let F be a cdf on R. Let F_{X_n} be the cdf of X_n $(n \geqslant 1)$. If $F_{X_n} \overset{w}{\rightarrow} F$, one says that $(X_n, n \geqslant 1)$ converges in law (or in distribution) to F. This is denoted by

$$X_n \overset{\mathscr{L}}{\longrightarrow} F \quad \text{or} \quad \mathscr{L}(X_n) \to F.$$

The three above definitions are basically the same. The best thing to do would be to ignore the last two definitions and only use the concept of weak convergence of cdf's. Indeed, Definitions D7 and D8 could mislead a beginner who could be tempted to see in the definition of convergence in law a property of the random variables themselves, whereas it is only a property of their cdf's. This is in sharp contrast with almost-sure convergence.

To illustrate this seemingly subtle point, let us consider the following situation. The random variables X_n $(n \geqslant 1)$ and X are supposed to be identically distributed with a common probability density f which is symmetric, i.e., $f(x) = f(-x)$ for all $x \in \mathbb{R}$. This implies in particular that $-X_n$ and $-X$ have the same pd, namely f.

Clearly, $X_n \overset{\mathscr{L}}{\longrightarrow} X$. If convergence in law could be treated in the same way as almost-sure convergence, one would expect the following consequence: $X_n - X \overset{\mathscr{L}}{\longrightarrow} 0$. But this is dramatically false in general. To see this, suppose that $(X_n, n \geqslant 1)$ and X are independent, then since X and $Y = -X$ have the same distribution, $X_n - X = X_n + Y$ has for pd the convolution $g(x) = \int_{-\infty}^{+\infty} f(x - z)f(z)\,dz$. Also, the corresponding cdf $G(x) = \int_{-\infty}^{x} g(u)\,du$, which is $P(X_n - X \leqslant x)$, does not in general converge weakly to $1_{\mathbb{R} - \{0\}}(x)$, which is the cdf of the null random variable.

E7 Exercise. Let $(X_n, n \geqslant 1)$ be a sequence of iid random variables uniformly distributed on $[0, 1]$. Define $Z_n = \min(X_1, \ldots, X_n)$. Show that $Z_n/n \overset{\mathscr{L}}{\longrightarrow} \mathscr{E}(1)$. [This notation means $Z_n/n \overset{\mathscr{L}}{\longrightarrow} F$ where $F(x) = 1 - e^{-x}$; F is the cdf corresponding to an exponential random variable of mean 1.]

EXAMPLE 3 (Integer-Valued Random Variables). For discrete random variables with integer values, the notion of convergence in distribution is trivial. The sequence of $\mathbb{N}$-valued random variables $(X_n, n \geq 1)$ converges in distribution to the $\mathbb{N}$-valued random variable X if

$$\lim_{n \uparrow \infty} P(X_n = i) = P(X = i) \qquad \text{for all} \quad i \in \mathbb{N}. \tag{10}$$

The reader can verify that Eq. (10) implies Eq. (9).

We can now see why, in the definition of convergence in law, the discontinuity points of the cdf play a special part (actually, do not play any part). If Eq. (9) were required to hold for *all* $x \in \mathbb{R}$, then defining

$$X_n \equiv a + \frac{1}{n}, \qquad X \equiv a,$$

we could not say that $X_n \xrightarrow{\mathscr{L}} X$, because $P(X_n \leq a) = 0$ does not converge toward $P(X \leq a) = 1$. To have a definition of convergence in law sufficiently rich, one must therefore exclude the discontinuity points (such as a in the above example) of X in Eq. (9).

E8 Exercise (Poisson's Law of Rare Events). Let $(X_n, n \geq 1)$ be a sequence of independent $\{0, 1\}$-valued random variables, with $P(X_n = 1) = p_n$ where the sequence $(p_n, n \geq 1)$ verifies $\lim_{n \uparrow \infty} n p_n = \lambda$ for some positive λ. Show that $X_1 + \cdots + X_n \xrightarrow{\mathscr{L}} \mathscr{P}(\lambda)$ where $\mathscr{P}(\lambda)$ is the Poisson distribution with mean λ.

We shall now state without proof the fundamental criterion of convergence in law. It will be stated in terms of weak convergence of cumulative distribution functions (Definition D6) but it will be used mainly for convergence in law.

T9 Theorem (Characteristic Function Criterion). *Let* $(F_n, n \geq 1)$ *be cdf's on* $\mathbb{R}$ *with respective characteristic functions* $(\phi_n, n \geq 1)$. *Suppose that*

$$\lim_{n \uparrow \infty} \phi_n(u) = \phi(u), \qquad \forall u \in \mathbb{R}, \tag{11}$$

for some $\phi : \mathbb{R} \to \mathbb{C}$ *that is continuous at* 0. *Then* $F_n \xrightarrow{w} F$ *for some cdf* F *which admits* ϕ *as characteristic function.*

In terms of random variables, the criterion reads as follows: *Let* $(X_n, n \geq 1)$ *be random variables with respective characteristic functions* $(\phi_n, n \geq 1)$.

If the sequence $(\phi_n, n \geq 1)$ *converges pointwise to some function* $\phi : \mathbb{R} \to \mathbb{C}$ *continuous at* 0, *then* ϕ *is a characteristic function* [*i.e.,* $\phi(u) = \int_{-\infty}^{+\infty} e^{iux} \, dF(x)$ *for some cdf* F] *and moreover,* $X_n \xrightarrow{\mathscr{L}} F$ (*in the sense of Definition D8*).

EXAMPLE 4 (DeMoivre–Laplace Central Limit Theorem). Let $(X_n, n \geq 1)$ be an iid sequence of $\{0, 1\}$-valued random variables with $P(X_n = 1) = p$ for some $p \in (0, 1)$. We shall see (in Exercise E9) that, letting $q = 1 - p$,

$$\frac{X_1 + \cdots + X_n - np}{\sqrt{npq}} \xrightarrow{\mathscr{L}} \mathscr{N}(0, 1), \tag{12a}$$

that is,

$$\lim_{n \uparrow \infty} P\left[\frac{X_1 + \cdots + X_n}{\sqrt{npq}} \leqslant x\right] = \frac{1}{\sqrt{2\pi}} \int_{-\infty}^{x} e^{-y^2/2} \, dy. \tag{12b}$$

E9 Exercise. Compute the cf ϕ_n of the random variable $(X_1 + \cdots + X_n - np)/\sqrt{npq}$ and show that $\lim_{n \uparrow \infty} \phi_n(u) = e^{-u^2/2}$ [the cf of $\mathscr{N}(0, 1)$]. Then use Theorem T9 to conclude that Eq. (12) holds.

E10 Exercise. Let $(X_n, n \geqslant 1)$ be a sequence of independent Cauchy random variables [recall that $\phi_{X_n}(u) = e^{-|u|}$]. Does the sequence $[(X_1 + \cdots + X_n)/n^2, n \geqslant 1]$ converge in law?

E11 Exercise. Let $(X_n, n \geqslant 1)$ be a sequence of random variables such that $P(X_n = k/n) = 1/n$, $1 \leqslant k \leqslant n$. Does it converge in law?

Case of Random Vectors. Definitions D6, D7, and D8 and Theorem T9 also apply to the case of random vectors, mutatis mutandis. For instance, a sequence $(X_n, n \geqslant 1)$ of k-dimensional random vectors is said to converge in law to the k-dimensional random vector X if and only if

$$\lim_{n \uparrow \infty} P(X_n \leqslant x) = P(X \leqslant x)$$

for all $x \in \mathbb{R}^k$ such that $P(X = x) = 0$. Recall that for a k-dimensional random vector Y, $P(Y \leqslant y)$ is a shorter notation for $P(Y_1 \leqslant y_1, \ldots, Y_k \leqslant y_k)$ where $Y = (Y_1, \ldots, Y_k)$ and $y = (y_1, \ldots, y_k)$.

The same notations as for the scalar case are used: $X_n \xrightarrow{\mathscr{L}} X$, $X_n \xrightarrow{\mathscr{L}} F$, etc. As for Theorem T9, it reads as follows: let $(X_n, n \geqslant 1)$ be a sequence of k-dimensional vectors with respective characteristic functions $(\phi_n, n \geqslant 1)$ such that $\lim_{n \uparrow \infty} \phi_n(u) = \phi(u)$ for all $u \in \mathbb{R}^k$, where ϕ is continuous at $0 \in \mathbb{R}^k$. Then ϕ is the cf of some cumulative distribution function F on $\mathbb{R}^k$ and $X_n \xrightarrow{\mathscr{L}} F$.

2.2. The Central Limit Theorem

The Laplace–DeMoivre result featured in the previous subsection [see Eq. (12a)] is an avatar of the second most celebrated result of Probability Theory after the strong law of large numbers: the central limit theorem, which will now be stated without proof.

T10 Theorem (Central Limit Theorem). *Let $(X_n, n \geqslant 1)$ be a sequence of iid random variables with common mean and variance μ and σ^2, respectively. Suppose moreover that $0 < \sigma^2 < \infty$. Then*

$$\frac{(X_1 + \cdots + X_n) - n\mu}{\sigma\sqrt{n}} \xrightarrow{\mathcal{L}} \mathcal{N}(0, 1), \tag{13a}$$

that is,

$$\lim_{n\uparrow\infty} P\left(\frac{X_1 + \cdots + X_n - n\mu}{\sigma\sqrt{n}} \leqslant x\right) = \frac{1}{\sqrt{2\pi}} \int_{-\infty}^{x} e^{-y^2/2}\, dy. \tag{13b}$$

Note that the mean of $Z_n = (X_1 + \cdots + X_n)/\sigma\sqrt{n}$ is 0 and that its variance is 1, i.e., Z_n is obtained by standardization of $X_1 + \cdots + X_n$.

A SKETCH OF PROOF FOR THE CENTRAL LIMIT THEOREM. Without loss of generality we can assume that $\mu = 0$. Denoting by ϕ the cf of X_1, the cf of $(X_1 + \cdots + X_n)/\sigma\sqrt{n}$ is $[\phi(u/\sigma\sqrt{n})]^n$. But $\phi(0) = 1$, $\phi'(0) = 0$ (since $EX = 0$) and $\phi''(0) = -\sigma^2$, and therefore, using the Taylor expansion of ϕ at 0,

$$\phi\left(\frac{u}{\sigma\sqrt{n}}\right) = 1 - \frac{1}{2n}u^2 + o\left(\frac{1}{n}\right).$$

Therefore

$$\lim_{n\uparrow\infty} \phi\left(\frac{u}{\sigma\sqrt{n}}\right)^n = \lim_{n\uparrow\infty}\left(1 - \frac{u^2}{2n}\right)^n = e^{-u^2/2}.$$

Theorem T9 yields the conclusion.   □

E12 Exercise. Use the central limit theorem to prove that

$$\lim_{n\uparrow\infty} e^{-n} \sum_{k=0}^{n} \frac{n^k}{k!} = \frac{1}{2}.$$

E13 Exercise. Let F be a cdf on $\mathbb{R}$ with mean 0 and variance 1. (i.e., $\int_{-\infty}^{+\infty} x\, dF(x) = 0$ and $\int_{-\infty}^{+\infty} x^2\, dF(x) = 1$). The following assumption on F is made: if X_1 and X_2 are two independent random variables with the same cdf F, then $(X_1 + X_2)/\sqrt{2}$ also admits F as cdf. Show (using the central limit theorem) that $F(x) = (1/\sqrt{2\pi})\int_{-\infty}^{x} e^{-y^2/2}\, dy$.

Application: Gauss' Theory of Errors. Suppose that you perform n independent experiments resulting in the data $X_1, \ldots, X_n$, which are independent and identically distributed random numbers of mean μ and variance σ^2 $(0 < \sigma^2 < \infty)$. This is typically the situation when you want to obtain experimentally the value μ of some observable quantity and when the experiments are subjected to independent errors, i.e., $X_n = \mu + \varepsilon_n$ where the random numbers $\varepsilon_1, \ldots, \varepsilon_n$ are iid, with mean 0 and variance σ^2.

The experimental value of the (unknown) quantity μ is the empirical mean $\bar{X}_n = (X_1 + \cdots + X_n)/n$. This "estimate" of μ has the interesting property: $\bar{X}_n \xrightarrow{\text{as}} \mu$ (strong law of large numbers). However, $\bar{X}_n$ is not exactly μ, and one would like to have some information about the distribution of X_n around its asymptotic value μ. For instance, one would like to have an estimate of

$$P(|\bar{X}_n - \mu| \geq a), \tag{14}$$

i.e., of the probability of making a mistake larger than a. If one does not know exactly the distribution of each X_n around its mean μ, it is impossible to compute Term (14) exactly. However, maybe the central limit theorem will help us in obtaining a sensible estimate of Term (14), if n is "large enough." To do this, observe that

$$P(|\bar{X}_n - \mu| \geq a) = P\left[\left|\frac{X_1 + \cdots + X_n - n\mu}{\sigma\sqrt{n}}\right| \geq \frac{a\sqrt{n}}{\sigma}\right].$$

Now suppose that n is large enough so that you can, in view of the central limit theorem, assimilate

$$P\left[\left|\frac{X_1 + \cdots + X_n - n\mu}{\sigma\sqrt{n}}\right| \geq b\right] \quad \text{to} \quad 1 - \frac{1}{\sqrt{2\pi}} \int_{-b}^{+b} e^{-y^2/2}\, dy.$$

You then have the approximate equality

$$P(|\bar{X}_n - \mu| \geq a) \simeq 1 - \frac{1}{\sqrt{2\pi}} \int_{-a\sqrt{n}/\sigma}^{+a\sqrt{n}/\sigma} e^{-y^2/2}\, dy. \tag{15}$$

The right-hand side of Eq. (15) is $2Q[a\sqrt{n}/\sigma]$ where $Q(x) = (1/\sqrt{2\pi}) \times \int_x^\infty e^{-y^2/2}\, dy$ is the "Gaussian tail" (Fig. 1).

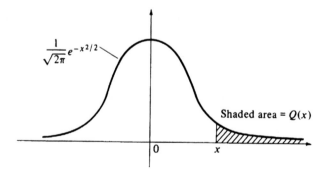

$\frac{1}{\sqrt{2\pi}} e^{-x^2/2}$

Shaded area $= Q(x)$

Figure 1. The Gaussian tail.

Statistical tables for the Gaussian tail are available in all statistical handbooks. Figure 2 is an excerpt of such a table that will be sufficient for our pedagogical purpose.

EXAMPLE 5. How many experiments would you perform so that the deviation from the exact mean would be less than 10% of the standard deviation with a probability larger than 0.99? Here we want

x	1.70	1.75	1.80	1.85	1.90	1.95	1.96	2.00	2.05	2.10
$1 - Q(x)$	0.955	0.960	0.964	0.968	0.971	0.974	0.975	0.977	0.980	0.982
x	2.15	2.20	2.25	2.30	2.326	2.35	2.40	2.45	2.50	2.55
$1 - Q(x)$	0.984	0.986	0.988	0.989	0.990	0.991	0.992	0.993	0.994	0.995
x	2.576	2.60	2.65	2.70	2.75	2.80	2.85	2.90	2.95	3.00
$1 - Q(x)$	0.995	0.995	0.996	0.997	0.997	0.997	0.998	0.998	0.998	0.999

Figure 2

$$P(|\bar{X}_n - \mu| \geqslant 0.1\sigma) < 0.01.$$

Applying Eq. (15) with $a = 0.1\sigma$, we see that the number of experiments n should be such that

$$2Q(\sqrt{n}/10) < 0.01.$$

From Fig. 2, we see that $n > (26)^2$ will do. Fortunately, in practice, much smaller n is enough. The estimate (15) is in fact not very tight and therefore not very useful in this particular situation.

The Notion of Confidence Interval. Suppose X is a random variable of unknown mean μ and known variance σ^2. One way of obtaining μ is to perform n identical independent experiments resulting in n iid random variables $X_1, \ldots, X_n$ of the same distribution as X and to approximate μ by the empirical mean $\bar{X}_n = (X_1 + \cdots + X_n)/n$. Now fix a confidence coefficient α (close to 1) and find the number a such that $2Q(a) = 1 - \alpha$. If n is large, we can use the central limit approximation

$$P\left(-a \leqslant \frac{\sqrt{n}}{\sigma}(\bar{X}_n - \mu) \leqslant a\right) \simeq 1 - \alpha.$$

But the event $-a \leqslant (\sqrt{n}/\sigma)(\bar{X}_n - \mu) \leqslant a$ is just $\bar{X}_n - a\sigma/\sqrt{n} \leqslant \mu \leqslant \bar{X}_n + a\sigma/\sqrt{n}$. For this reason, the (random) interval $[\bar{X}_n - (a\sigma/\sqrt{n}), \bar{X}_n + (a\sigma/\sqrt{n})]$ is called a confidence interval for the mean μ, relative to the level of confidence α. You will trust that μ is in this interval with a probability α.

EXAMPLE 6 (Testing a Coin). Suppose that $X = 0$ or 1 (e.g., a coin, where heads $= 1$), and that $P(X = 1) = p$ is unknown. The situation is almost the same as above, with $\mu = p$, except that σ is unknown. In this case, one should take an upperbound for σ. Here $\sigma = \sqrt{pq} < \frac{1}{2}$. Therefore, the confidence interval for the level of confidence $\alpha = 0.99$ ($a = 2.6$) will be $[\bar{X}_n - (1.3/\sqrt{n}), \bar{X}_n + (1.3/\sqrt{n})]$. For instance, if you obtain 4.950 heads in 10.000 tosses, you can say that the bias p of the coin is between $0.495 - 0.013 = 0.482$ and $0.495 + 0.013 = 0.508$ with probability 0.99. This is compatible with a fair coin ($p = \frac{1}{2}$).

3. The Hierarchy of Convergences

So far we have seen three notions of convergence: convergence in probability, convergence in law, and almost-sure convergence. We shall soon add the definition of convergence in quadratic mean to our collection. But then, what are the relations between these apparently different types of convergence? It turns out that convergence in probability, convergence in quadratic mean, and almost-sure convergence, although different, are closely connected, whereas convergence in law has an altogether different status, as we already mentioned.

3.1. Almost-Sure Convergence Versus Convergence in Probability

Let X_n, $n \geq 1$ and X be random variables. Recall that $X_n \xrightarrow{\text{Pr}} X$ means that $\lim_{n \uparrow \infty} P(|X_n - X| \geq \varepsilon) = 0$ for all $\varepsilon > 0$, and that it reads: $(X_n, n \geq 1)$ converges in probability to X.

T11 Theorem.

(a) *If* $X_n \xrightarrow{\text{as}} X$, *then* $X_n \xrightarrow{\text{Pr}} X$.
(b) *If* $X_n \xrightarrow{\text{Pr}} X$, *then there exists a strictly increasing sequence of integers* $(n_k, k \geq 1)$ *such that* $X_{n_k} \xrightarrow[k \uparrow \infty]{\text{as}} X$. *In other words, one can extract from a sequence converging in probability to some random variable X a subsequence converging almost surely to the same random variable X.*

PROOF OF THEOREM T11.

(a) $X_n \xrightarrow{\text{as}} X$ is equivalent to $|X_n - X| \xrightarrow{\text{as}} 0$, and this is in turn equivalent (by Theorem T4) to $P(\overline{\lim}\{|X_n - X| \geq \varepsilon\}) = 0$ for all $\varepsilon > 0$. But $P(\overline{\lim}\{|X_n - X| \geq \varepsilon\}) = \lim_{n \uparrow \infty} \downarrow P(\bigcup_{k \geq n}\{|X_k - X| \geq \varepsilon\})$, and $P(\bigcup_{k \geq n} \cdot \{|X_k - X| \geq \varepsilon\}) \geq P(|X_n - X| \geq \varepsilon)$. Therefore, $\lim_{n \uparrow \infty} P(|X_n - X| \geq \varepsilon) = 0$, for all $\varepsilon \geq 0$.
(b) Define recursively the sequence $(n_k, k \geq 1)$ as follows: n_{k+1} is any index $n > n_k$ such that $P(|X_{n_{k+1}} - X| \geq 1/k) \leq 1/2^k$. If $\lim_{n \uparrow \infty} P(|X_n - X| \geq \varepsilon) = 0$ such a sequence is constructible, and by Exercise E4, $X_{n_k} \xrightarrow[k \uparrow \infty]{\text{as}} X$. $\square$

A Counterexample. There exist sequences of random variables that converge in probability but not almost surely. Consider, for instance, $(X_n, n \geq 1)$, an independent sequence with $P(X_n = 1) = 1 - P(X_n = 0) = p_n = 1 - q_n$, where

$$\lim_{n \uparrow \infty} p_n = 0, \qquad \sum_{n=1}^{\infty} p_n = \infty. \qquad (16)$$

It was shown in Exercise E2 that such a sequence does not converge to 0 because $\sum_{n=1}^{\infty} p_n = \infty$. Similarly, it does not converge to 1 since $\sum_{n=1}^{\infty} q_n = \infty$.

Therefore, it does not converge at all in the almost-sure sense. But for any $\varepsilon > 0$, $P(X_n \geq \varepsilon) \leq p_n$, and therefore $X_n \xrightarrow{\text{Pr}} 0$.

3.2. Convergence in the Quadratic Mean

D12 Definition. A sequence of random variables $(X_n, n \geq 1)$ such that $E[|X_n|^2] < \infty$ $(n \geq 1)$ is said to converge in the quadratic mean to a random variable X such that $E[|X|^2] < \infty$ if

$$\lim_{n \uparrow \infty} E[|X_n - X|^2] = 0. \tag{17}$$

This is denoted by $X_n \xrightarrow{\text{qm}} X$.

This type of convergence is not as intuitively appealing as convergence in law or almost-sure convergence. It plays a technical role in Probability Theory, especially in situations where only "second-order data" are available, i.e., when all the probabilistic information about a given sequence $(X_n, n \geq 1)$ of random variables consists of the mean function $m_X : \mathbb{N}_+ \to \mathbb{R}$, where $m_X(n) = EX_n$, and the autocorrelation function $R_X : \mathbb{N}_+^2 \to \mathbb{R}$ where $R(k, n) = E[(X_k - m_X(k))(X_n - m_X(n))]$. In such a case, one cannot compute in general the cf $\phi_n(u) = E[e^{iuX_n}]$ or evaluate quantities such as $P(X_n \geq \varepsilon)$, which intervene in the criteria of convergence in law and almost-sure convergence, respectively. On the contrary, second-order data are enough to determine whether $(X_n, n \geq 1)$ converges in the quadratic mean or not. Indeed, there is a criterion of the Cauchy type (which we shall not prove) involving only m_X and R_X:

T13 Theorem. *A sequence of random variables $(X_n, n \geq 1)$ such that $E[|X_n|^2] < \infty$ $(n \geq 1)$ converges in the quadratic mean to some random variable X if and only if*

$$\lim_{m, n \uparrow \infty} E[|X_m - X_n|^2] = 0. \tag{18}$$

Markov's inequality $P(|X_n - X| \geq \varepsilon) \leq E[|X_n - X|^2]/\varepsilon^2$ shows that if $X_n \xrightarrow{\text{qm}} X$, then $X_n \xrightarrow{\text{Pr}} X$. Therefore, in the hierarchy of convergences, qm is above Pr. But we can say more:

T14 Theorem. *Convergence in probability implies convergence in law (and therefore convergence in the quadratic mean implies convergence in law).*

PROOF OF THEOREM T14. Let $(X_n, n \geq 1)$ be a sequence of random variables such that $X_n \xrightarrow{\text{Pr}} X$ for some random variable X, i.e., $\lim_{n \uparrow \infty} P(|X_n - X| \geq \varepsilon)$ for all $\varepsilon > 0$. Therefore, for all $\varepsilon > 0$ and all $\delta > 0$, there exists N such that

$n \geqslant N$ implies

$$P(|X_n - X| \geqslant \varepsilon) \leqslant \delta \qquad (*)$$

and therefore (see Exercise E14), for all $x \in \mathbb{R}$,

$$P(X \leqslant x - \varepsilon) - \delta \leqslant P(X_n \leqslant x) \leqslant P(X \leqslant x + \varepsilon) + \delta. \qquad (**)$$

If x is a continuity point of X, i.e., $P(X = x) = 0$, ε can be chosen such that

$$P(X \leqslant x - \varepsilon) \geqslant P(X \leqslant x) - \delta, \qquad P(X \leqslant x + \varepsilon) \leqslant P(X \leqslant x) + \delta.$$

Therefore if $x \in \mathbb{R}$ is such that $P(X = x) = 0$, for all $\delta > 0$ there exists N such that

$$P(X \leqslant x) - 2\delta \leqslant P(X_n \leqslant x) \leqslant P(X \leqslant x) + 2\delta \qquad (n \geqslant N).$$

Thus, $\lim_{n \uparrow \infty} P(X_n \leqslant x) = P(X \leqslant x)$, qed. $\qquad \square$

E14 Exercise. Prove that Term $(*)$ implies Term $(**)$.

E15 Exercise. Prove that if $X_n \overset{\text{as}}{\longrightarrow} X$ and $|X_n| \leqslant Y$ with $E[Y^2] < \infty$, then $X_n \overset{\text{qm}}{\longrightarrow} X$.

Stability of the Gaussian Character. We have seen that Gaussian vectors remain Gaussian after an affine transformation (in fact, this was part of Definition D1 of Chapter 4). Gaussian random variables have another interesting stability property.

T15 Theorem. *If $(X_n, n \geqslant 1)$ is a sequence of Gaussian random variables such that $X_n \overset{\text{qm}}{\longrightarrow} X$, then X is also Gaussian.*

PROOF OF THEOREM T15. From Theorem T14, $X_n \overset{\mathscr{L}}{\longrightarrow} X$ and therefore, from Theorem T9,

$$\phi_X(u) = \lim_{n \uparrow \infty} \phi_{X_n}(u)$$

where ϕ_X and ϕ_{X_n} are the characteristic functions of X and X_n, respectively. But

$$\phi_{X_n}(u) = e^{E[X_n]u - (1/2)(E[X_n^2] - E[X_n]^2)u^2}.$$

But (see Exercise E16), $\lim E[X_n] = E[X] = m_X$ and $\lim E[X_n^2] = E[X^2] = \sigma_X^2 + m_X^2$. Therefore,

$$\phi_X(u) = e^{im_X u - (1/2)\sigma_X^2 u^2}. \qquad \square$$

E16 Exercise. Show that if $X_n \overset{\text{qm}}{\longrightarrow} X$ and $Y_n \overset{\text{qm}}{\longrightarrow} Y$, then $\lim_{m,n \uparrow \infty} E[X_n Y_m] = E[XY]$, and in particular, $\lim_{n \uparrow \infty} E[X_n^2] = E[X^2]$ and $\lim_{n \uparrow \infty} E[X_n] = E[X]$.

3.3. Convergence in Law in the Hierarchy of Convergences

Convergence in law of a sequence $(X_n, n \geq 1)$ is relative to the cdf's of the X_n's. This means that we can talk about convergence in law even when the X_n's are not defined on the same probability space! Of course, this may seem to be only a mathematical subtlety, and indeed it is one. But this explains why we cannot issue a theorem stating that convergence in law implies another type of convergence, unless we require that the X_n's are defined on the same probability space. But even if we do, the situation of convergence in law is not brilliant.

E17 Exercise. Let $(X_n, n \geq 1)$ be a sequence of random variables defined on the same probability space, and suppose that for some constant c, $X_n \xrightarrow{\mathscr{L}} c$ (meaning $\lim_{n \uparrow \infty} P(X_n \leq x) = 0$ if $x < c$ and 1 if $x > c$). Show that $X_n \xrightarrow{\text{Pr}} c$.

There is a result due to Skorokhod, which says that "if you change probability space, convergence in law is as good as almost-sure convergence." More precisely, let $(X_n, n \geq 1)$ be a sequence of random variables defined on $(\Omega, \mathscr{F}, P)$ together with some other random variable X and suppose that $X_n \xrightarrow{\mathscr{L}} X$. Then one can exhibit a probability space $(\tilde{\Omega}, \tilde{\mathscr{F}}, \tilde{P})$ and random variables $\tilde{X}_n$ $(n \geq 1)$ and $\tilde{X}$ such that

$$\begin{cases} \tilde{P}(\tilde{X}_n \leq x) = P(X_n \leq x) & (n \geq 1, x \in \mathbb{R}) \\ \tilde{P}(\tilde{X} \leq x) = P(X \leq x) & (x \in \mathbb{R}) \end{cases} \tag{19}$$

and

$$\tilde{X}_n \xrightarrow{\tilde{P}\text{-as}} X. \tag{20}$$

The "converse" of Skorokhod's result is true without changing probability spaces.

T16 Theorem. *Let $(X_n, n \geq 1)$ and X be random variables defined on $(\Omega, \mathscr{F}, P)$ such that $X_n \xrightarrow{\text{as}} X$. Then $X_n \xrightarrow{\mathscr{L}} X$.*

PROOF OF THEOREM T16. If $X_n \xrightarrow{\text{as}} X$, then $e^{iuX_n} \xrightarrow{\text{as}} e^{iuX}$. Now $|e^{iuX_n}| \leq 1$, and therefore (Lebesgue's dominated convergence theorem), $\lim_{n \uparrow \infty} E[e^{iuX_n}] = E[e^{iuX}]$. Hence, (Theorem T9), $X_n \xrightarrow{\mathscr{L}} X$. $\qquad\square$

3.4. The Hierarchical Tableau

Here $(X_n, n \geq 1)$ is a sequence of random variables defined on $(\Omega, \mathscr{F}, P)$, together with X (Fig. 3). We shall now illustrate some of the results from the above hierarchical tableau.

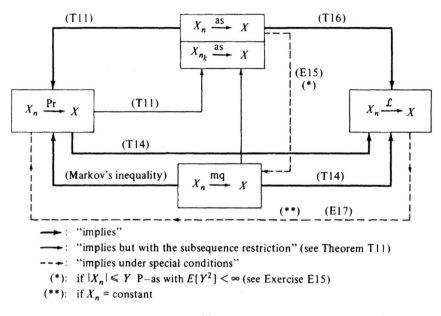

: "implies"

: "implies but with the subsequence restriction" (see Theorem T11)

- - →: "implies under special conditions"

(*): if $|X_n| \leqslant Y$ P$-$as with $E[Y^2] < \infty$ (see Exercise E15)

(**): if X_n = constant

Figure 3

Cauchy: A Bag Full of Examples. Let $(X_n, n \geqslant 1)$ be an iid sequence of Cauchy random variables defined on $(\Omega, \mathscr{F}, P)$. In particular, the cf of any of these random variables is $e^{-|u|}$. Define

$$Y_n = \frac{X_1 + \cdots + X_n}{\sqrt{n}} \qquad (21)$$

$$Z_n = \frac{X_1 + \cdots + X_n}{n} \qquad (22)$$

$$W_n = \frac{X_1 + \cdots + X_n}{n^2}. \qquad (23)$$

We are going to examine the convergence of the three sequences $(Y_n, n \geqslant 1)$, $(Z_n, n \geqslant 1)$, and $(W_n, n \geqslant 1)$. To discuss convergence in law we need the characteristic functions

$$\phi_{Y_n}(u) = e^{-\sqrt{n}|u|} \qquad (24)$$

$$\phi_{Z_n}(u) = e^{-|u|} \qquad (25)$$

$$\phi_{W_n}(u) = e^{-|u|/n}. \qquad (26)$$

Since

$$\lim_{n \uparrow \infty} \phi_{Y_n}(u) = \phi(u) = \begin{cases} 1 & \text{if } u = 0 \\ 0 & \text{if } u \neq 0 \end{cases}$$

is not continuous at $u = 0$, it cannot be a cf, and therefore (Theorem T9) $(Y_n, n \geq 1)$ does not converge in law. It cannot converge almost surely to some random variable X, or otherwise it would converge in law to the same X (see Theorem T16). By Theorem T14, $(Y_n, n \geq 1)$ cannot converge in probability or in the quadratic mean.

By Theorem T9, and Eq. (25), $Z_n \xrightarrow{\mathscr{L}}$ Cauchy. Recall that if X_1 is Cauchy, $E|X_1| = \infty$, and therefore Kolmogorov's strong law of large numbers is not applicable. In fact, even if we make the "natural" choice for the mean of X_1 (which we must insist does not exist), i.e., "$m_{X_1} = 0$", we do not have $Z_n \xrightarrow{\text{as}} 0$ because (by Theorems T11 and T14) this would imply $Z_n \xrightarrow{\mathscr{L}} 0$, in contradiction with $Z_n \xrightarrow{\mathscr{L}}$ Cauchy.

Since $\lim_{n \uparrow \infty} \phi_{W_n}(u) = 1$, which is the cf of a random variable equal to 0, we have, by Theorem T9, $W_n \xrightarrow{\mathscr{L}} 0$. By Exercise E17, $W_n \xrightarrow{\text{Pr}} 0$. By Theorem T11, there exists a subsequence $(W_{n_k}, k \geq 1)$ such that $W_{n_k} \xrightarrow{\text{as}} 0$. The notion of convergence in quadratic mean is not applicable to $(W_n, n \geq 1)$ since $E(|W_n|^2) = \infty$ for all $n \geq 1$.

Illustration 8. A Statistical Procedure: The Chi-Square Test

The theory of convergence in law finds a priviledged domain of application in Statistics. Although the chi-square test is only one of the numerous statistical procedures based on a result of convergence in law, it is quite representative of the way statisticians think and operate when they wish to identify a probabilistic model, or at least to validate some hypothesis concerning this model.

The problem of concern to us in this Illustration is the following. You observe n iid random variables $X_1, \ldots, X_n$ taking their values in a finite set $E = \{1, \ldots, k\}$, and you want to know to what extent you can believe in the hypothesis

$$P(X_1 = i) = p_i \qquad (1 \leq i \leq k) \tag{27}$$

where $p = (p_1, \ldots, p_k)$ is a given probability distribution over E.

EXAMPLE 1. A dice must be tested for unbiasedness. Here $E = \{1, \ldots, 6\}$ and $p_i = 1/6 \, (1 \leq i \leq 6)$.

EXAMPLE 2. One observes n iid random variables $Y_1, \ldots, Y_n$ and wishes to test the validity of hypothesis

$$P(Y \leq x) = F(x)$$

where $Y = Y_1$ and F is some probability distribution function over $\mathbb{R}$. This problem can be converted to the basic problem by considering a partition

Illustration 8. A Statistical Procedure: The Chi-Square Test   181

$\{\Delta_i, 1 \leqslant i \leqslant k\}$ of $\mathbb{R}$ and letting for each $i\,(1 \leqslant i \leqslant k)$

$$X_i = \begin{cases} i & \text{if } Y_i \in \Delta_i \\ 0 & \text{otherwise} \end{cases}$$

and

$$p_i = P(Y_1 \in \Delta_i).$$

If hypothesis (27) is true, one expects, in view of the law of large numbers, that for each $i\,(1 \leqslant i \leqslant k)$ the empirical frequency

$$\frac{1}{n}N_i^{(n)} = \frac{1}{n}\sum_{j=1}^{n}1_{\{x_j=i\}} \tag{28}$$

is close to p_i.

The chi-square test proposes to measure the deviation to the law of large numbers by

$$\boxed{T_n = \sum_{i=1}^{k} \frac{(N_i^{(n)} - np_i)^2}{np_i}.} \tag{29}$$

The random variable T_n is called the *Pearson statistics* of order n relative to the test of hypothesis $p = (p_1, \ldots, p_k)$. If p is the correct hypothesis, then, as we shall soon see, T_n converges in law to a chi-square distribution. Otherwise T_n goes to ∞. Therefore, a theoretically reasonable test procedure is the following. First fix a (small) number $\alpha > 0$, called the *level of the test*. Let P be the probability under which Eq. (27) is true, and let $t_0(\alpha)$ be the smallest s such that $P(T_n \geqslant s) \leqslant \alpha$ (Fig. 4). If the experimental value t of T_n verifies $t \leqslant t_0(\alpha)$, hypothesis $p = (p_1, \ldots, p_k)$ is accepted, otherwise it is rejected.

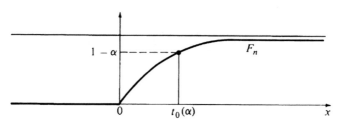

F_n is the distribution of T_n under hypothesis (27)

Figure 4. The level of the test and the acceptation threshold.

Although the principle of this test is intuitively appealing, its implementation is too difficult because it requires for the computation of $t_0(\alpha)$ a knowledge of the distribution function of T_n. Such a distribution function can in principle be computed, but it depends on n and on $p = (p_1, \ldots, p_k)$. Since statisticians are working with statistical tables, the procedure is not applicable as such.

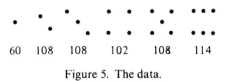

60   108   108   102   108   114

Figure 5. The data.

The chi-square test proposes to approximate $P(T_n \leqslant x)$ by $\lim_{n \uparrow \infty} P(T_n \leqslant x)$, a reasonable idea if the size n of sample $(X_1, \ldots, X_n)$ is large.

The interesting feature about the Pearson statistics is that

$$\lim_{n \uparrow \infty} P(T_n \leqslant x) = \frac{1}{2^{k/2} \Gamma\left(\dfrac{k}{2}\right)} \int_0^x y^{k/2-1} e^{-y/2} \, dy \qquad (x \geqslant 0) \, , \tag{30}$$

and therefore the limiting distribution does not depend on $p = (p_1, \ldots, p_k)$, and of course does not depend on n. The distribution function in the right-hand side of Eq. (30) is just the chi-square distribution with $k - 1$ degrees of freedom. The implementation of the chi-square test only requires the tabulation of this distribution function for various k.

EXAMPLE 1 (continued). The dice has been tossed $n = 600$ times, and it was recorded as in Fig. 5. The experimental value of T_n is

$$t = \frac{(60 - 100)^2 + (108 - 100)^2 + (108 - 100)^2 + (102 - 100)^2 + (108 - 100)^2 + (114 - 100)^2}{100}$$

$$= 19.92.$$

We want to be sure that the dice is unbiased, so we take a small value of the level, say $\alpha = 0.005$. If we call F_5 the chi-square distribution function with $6 - 1 = 5$ degrees of freedom, we find, using a table of F_5, that $t_0(\alpha) = 0.412$ $[1 - F_5(0.412) = 0.005]$. Since $19.92 > 0.412$, we must reject the hypothesis $p_i = (1/6) \, (1 \leqslant i \leqslant 6)$.

Exercise. One extracts 250 digits at ramdom from a table of ramdom digits. The result is

digit	0	1	2	3	4	5	6	7	8	9
number of occurrences	32	22	23	31	21	23	28	25	18	27

Test the equiprobability hypothesis with a chi-square test at level 0.1.

It now remains to prove Eq. (30), that is,

$$T_n \xrightarrow{\mathscr{L}} \chi^2_{k-1}. \tag{31}$$

Illustration 8. A Statistical Procedure: The Chi-Square Test   183

To do this, first consider the vector

$$Z_n = \left(\frac{N_1^{(n)} - np_1}{\sqrt{np_1}}, \ldots, \frac{N_k^{(n)} - np_k}{\sqrt{np_k}} \right). \tag{32}$$

We will show that it converges in law toward a Gaussian distribution admitting the characteristic function

$$\phi_Z(u) = \exp -(1/2) \left\{ \sum_{j=1}^{k} u_j^2 + \left(\sum_{j=1}^{k} \sqrt{p_j} u_j \right)^2 \right\}. \tag{33}$$

In particular, denoting by $\|x\|$ the euclidean norm of a vector $x \in \mathbb{R}^k$,

$$\lim_{n \uparrow \infty} P(T_n \leqslant x) = \lim_{n \uparrow \infty} P(\|Z_n\|^2 \leqslant x) = P(\|Z\|^2 \leqslant x). \tag{34}$$

We will then prove the existence of a standard Gaussian vector Y of dimension $k - 1$ such that

$$P(\|Z\|^2 \leqslant x) = P(\|Y\|^2 \leqslant x). \tag{35}$$

The proof will then be complete if we remember that the squared norm of a standard Gaussian vector of dimension l admits a chi-square distribution with l degrees of freedom.

Let us now prove the convergence in law of Z_n defined by Eq. (32) to the distribution F admitting the characteristic function ϕ_Z given by Eq. (33). We start by defining

$$M_j^{(i)} = 1_{\{x_i = j\}}$$

and

$$Y_i = \left(\frac{M_1^{(i)} - p_1}{\sqrt{np_1}}, \ldots, \frac{M_k^{(i)} - p_k}{\sqrt{np_k}} \right).$$

We have

$$Z_n = \sum_{i=1}^{n} Y_i,$$

and therefore, since the vectors $Y_1, \ldots, Y_n$ are independent and identically distributed,

$$\phi_{Z_n}(u) = (\phi_{Y_1}(u))^n.$$

Taking into account that for one index j, $M_1^{(j)} = 1$, and for all other indices $k \neq j$, $M_1^{(k)} = 0$,

$$\phi_{Y_1}(u) = Ee^{i \sum_{j=1}^{k} u_j[(M_j^{(1)} - p_j)/\sqrt{np_j}]} = \sum_{j=1}^{k} p_j e^{-i \sum_{l=1}^{k} (p_l u_l/\sqrt{np_l})} e^{i(u_j/\sqrt{np_j})}$$

$$= e^{-i \sum_{l=1}^{k} (p_l u_l/\sqrt{np_l})} \sum_{j=1}^{k} p_j e^{iu_j/\sqrt{np_j}}.$$

Therefore,

$$\log \phi_{Z_n}(u) = n \log \phi_{Y_1}(u)$$

$$= -in \sum_{l=1}^{k} \frac{p_l u_l}{\sqrt{np_l}} + n \log \sum_{j=1}^{k} p_j e^{iu_j/\sqrt{np_j}}.$$

Now, since $\sum_{j=1}^{k} p_j = 1$,

$$\log \left(\sum_{j=1}^{k} p_j e^{iu_j/\sqrt{np_j}} \right) = \log \left(1 + \sum_{j=1}^{k} (p_j e^{iu_j/\sqrt{np_j}} - 1) \right).$$

From the last two equalities, one obtains the development

$$\log \phi_{Z_n}(u) = \frac{1}{2} \sum_{j=1}^{k} u_j^2 + \frac{1}{2} \left(\sum_{j=1}^{k} \sqrt{p_j} u_j \right)^2 + o(1)$$

from which the announced convergence follows. It now remains to show that if a k-vector Z admits ϕ_Z given by Eq. (33) as its characteristic function, then $\|Z\|^2$ is distributed according to a chi-square distribution with $k - 1$ degrees of freedom. For this, we perform a change of orthonormal basis in $\mathbb{R}^k$, which relates the ancient coordinates u of a joint M to the new coordinates v according to

$$v = Au,$$

where A is a unitary $(k \times k)$ matrix $[AA' = A'A = I$ where $A' = $ transpose of A, and $I = (k \times k)$ identity matrix]. Therefore,

$$\|u\|^2 = \|v\|^2. \tag{36}$$

We shall select A in an ad hoc manner. Consider the hyperplane (Π) of $\mathbb{R}^k$ admitting the Cartesian representation

$$\sum_{j=1}^{k} \sqrt{p_j} u_j = 0,$$

and choose the kth axis of the new referential in such a way that v_k is the distance from M to (Π), i.e.,

$$v_k = \frac{\sum_{j=1}^{k} \sqrt{p_j} u_j}{\sum_{j=1}^{k} (\sqrt{p_j})^2} = \sum_{j=1}^{k} \sqrt{p_j} u_j. \tag{37}$$

The other axes of the new referential can then be chosen arbitrarily, under the constraint that the new referential is orthonormal.

From Eqs. (36) and (37) we obtain

$$\frac{1}{2} \sum_{j=1}^{k} u_j^2 - \frac{1}{2} \left(\sum_{j=1}^{k} \sqrt{p_j} u_j \right)^2 = \frac{1}{2} \sum_{j=1}^{k-1} v_j^2.$$

Therefore,

Illustration 9. Introduction to Signal Theory: Filtering 185

$$\phi_Z(A'v) = e^{-(1/2)\sum_{j=1}^{k-1} v_j^2}$$

(use the fact that $A' = A^{-1}$ when A is unitary). Defining

$$Y = AZ,$$

we find that

$$\phi_Y(v) = \phi_Z(A'v) = e^{-(1/2)\sum_{j=1}^{k-1} v_j^2}.$$

Hence, $Y = (Y_1, \ldots, Y_{k-1}, Y_k)$ is a degenerate Gaussian vector with $Y_k = 0$, P-as, and $(Y_1, \ldots, Y_{k-1})$ is a standard $(k-1)$ dimensional random vector. Also,

$$\|Z\|^2 = Z'IZ = Z'A'AZ = Y'Y = \|Y\|^2,$$

and therefore

$$\|Z\|^2 = \sum_{j=1}^{k-1} Y_j^2.$$

Illustration 9. Introduction to Signal Theory: Filtering

Linear Filtering of a Signal. Let $[X(t), t \in \mathbb{R}]$ be a *random signal*, that is, a real valued stochastic process (see Section 4, Chapter 4), with mean 0 and continuous autocorrelation function Γ_X:

$$E[X(t)] = 0, \qquad E[X(t)X(s)] = \Gamma_X(t, s). \tag{38}$$

Let now $h : \mathbb{R} \to \mathbb{R}$ be a piecewise continuous function null outside some bounded interval $[a, b]$. The operation transforming the random signal $[X(t), t \in \mathbb{R}]$ into the random signal $(Y(t), t \in \mathbb{R})$ defined by

$$Y(t) = \int_{-\infty}^{+\infty} h(t - s)X(s)\, ds \tag{39}$$

is called (linear homogeneous) *filtering* (Fig. 6). The function h is called the *impulse response* of the corresponding filter because if one enters the function

$$x_\varepsilon(t) = \begin{cases} \dfrac{1}{\varepsilon} & \text{if } t \in [0, \varepsilon] \\ 0 & \text{otherwise} \end{cases}$$

then the output

$$y_\varepsilon(t) = \int_{-\infty}^{+\infty} h(t - s)x_\varepsilon(s)\, ds = \frac{1}{\varepsilon}\int_0^\varepsilon h(t - s)\, ds$$

is nearly equal to $h(t)$ when ε is small (Fig. 7).

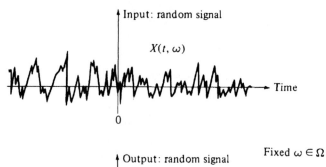

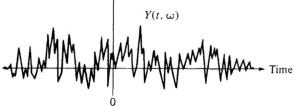

Figure 6. Random input and random output.

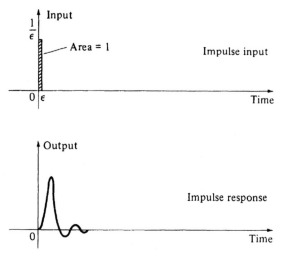

Figure 7. Impulse response.

A function such as x_ε, which is null outside $[0,\varepsilon]$ and of total area $\int_{-\infty}^{+\infty} x_\varepsilon(t)\,dt = 1$, is called a unit pulse at 0 when ε is small. This explains the appelation of h. The filtering operation is called linear because it is *linear in the input* $(x(t), t \in \mathbb{R})$, i.e.,

Illustration 9. Introduction to Signal Theory: Filtering   187

$$\int_{-\infty}^{+\infty} h(r - s)[\lambda_1 x_1(s) + \lambda_2 x_2(s)] \, ds$$

$$= \lambda_1 \int_{-\infty}^{+\infty} h(r - s)x_1(s) \, ds + \lambda_2 \int_{-\infty}^{+\infty} h(r - s)x_2(s) \, ds$$

for all $\lambda_1, \lambda_2 \in \mathbb{R}$, and all continuous functions $x_1, x_2 : \mathbb{R} \to \mathbb{R}$. It is said to be homogeneous because it is *homogeneous in time*, i.e., if

$$y(t) = \int_{-\infty}^{+\infty} h(t - s)x(s) \, ds$$

then

$$\int_{-\infty}^{+\infty} h(t - s)x(s - T) \, ds = y(t - T).$$

Thus, if the input x is delayed by T time units, the output is delayed by the same amount of time.

Many physical transformations can be described in terms of filtering. This is why filtering is a fundamental notion of signal theory.

Autocorrelation Function of a Filtered Random Signal. Returning to the output signal defined by Eq. (39), we shall prove that its mean is null, i.e.,

$$E[Y(t)] = 0 \tag{40}$$

and that its autocorrelation function $\Gamma_Y(t, s) = E[Y(t)Y(s)]$ is given by

$$\Gamma_Y(t, s) = \int_{-\infty}^{+\infty} \int_{-\infty}^{+\infty} h(t - u)h(t - v)\Gamma_X(u, v) \, du \, dv. \tag{41}$$

Since Eqs. (40) and (41) are particular cases of

$$E\left[\int_{-\infty}^{+\infty} f(u)X(u) \, du \right] = 0 \tag{42}$$

$$E\left[\int_{-\infty}^{+\infty} f(u)X(u) \, du \int_{-\infty}^{+\infty} g(v)X(v) \, dv \right] = \int_{-\infty}^{+\infty} \int_{-\infty}^{+\infty} f(u)g(v)\Gamma_X(u, v) \, du \, dv \tag{43}$$

where $f, g : \mathbb{R} \to \mathbb{R}$ are continuous functions, null outside some bounded interval (say $[0, 1]$ for the sake of notational ease), we will prove Eqs. (42) and (43). It will be assumed that for each $\omega \in r$, the trajectories $t \to X(t, \omega)$ are continuous, except on a finite set $A(\omega) \in [0, 1]$ (this is not a restriction in practice).

Define for each $n \geq 1$

$$\begin{cases} X_n = \dfrac{1}{2^n} \sum_{i=0}^{2^n-1} f\left(\dfrac{i}{2^n}\right) X\left(\dfrac{i}{2^n}\right) \\[3mm] Y_m = \dfrac{1}{2^m} \sum_{j=0}^{2^m-1} g\left(\dfrac{j}{2^m}\right) X\left(\dfrac{j}{2^m}\right). \end{cases} \tag{44}$$

By definition of Riemann's integral

$$X_n \xrightarrow{\text{as}} X = \int_0^1 f(s)X(s)\,ds, \qquad Y_n \xrightarrow{\text{as}} Y = \int_0^1 g(s)X(s)\,ds. \tag{45}$$

Suppose that we can prove that the convergences $X_n \to X$ and $Y_n \to Y$ also hold in the quadratic mean sense. Then Eqs. (42) and (43) follow almost immediately from the results of Exercise E16 (continuity of the scalar product). To see this, observe that for all $n \geq 1$, $m \geq 1$

$$E[X_n Y_m] = E\left[\sum_{i=0}^{2^n-1}\sum_{j=0}^{2^m-1} f\left(\frac{i}{2^n}\right)g\left(\frac{j}{2^m}\right)X\left(\frac{i}{2^n}\right)X\left(\frac{j}{2^m}\right)\right]\frac{1}{2^n}\frac{1}{2^m}$$

$$= \sum_{i=0}^{2^n-1}\sum_{j=0}^{2^m-1} f\left(\frac{i}{2^n}\right)g\left(\frac{j}{2^m}\right)E\left[X\left(\frac{i}{2^n}\right)X\left(\frac{j}{2^m}\right)\right]\frac{1}{2^n}\frac{1}{2^m}$$

$$= \sum_{i=0}^{2^n-1}\sum_{j=0}^{2^m-1} f\left(\frac{i}{2^n}\right)g\left(\frac{j}{2^m}\right)\Gamma_X\left(\frac{i}{2^n},\frac{j}{2^m}\right)\frac{1}{2^n}\frac{1}{2^m}.$$

Therefore, since f, g, and Γ_X are continuous,

$$\lim_{n,m\uparrow\infty} E[X_n Y_m] = \int_0^1\int_0^1 f(u)g(v)\Gamma_X(u,v)\,du\,dv. \tag{46}$$

On the other hand (by Exercise E16), if $X_n \xrightarrow{\text{qm}} X$, $Y_n \xrightarrow{\text{qm}} Y$, then

$$\lim_{n,m\uparrow\infty} E[X_n Y_m] = E[XY] = E\left[\int_0^1 f(u)X(u)\,du \int_0^1 g(v)X(v)\,dv\right]$$

and therefore Eq. (43) is proven. As for Eq. (42), it suffices to observe that for all $n \geq 1$

$$E[X_n] = \sum_{i=0}^{2^n-1} f\left(\frac{i}{2^n}\right)E\left[X\left(\frac{i}{2^n}\right)\right] = 0,$$

by hypothesis (38), and therefore (by Exercise E16, observing that the constant random variable 1 converges in quadratic mean to 1)

$$E\left[\int_0^1 f(u)X(u)\,du\right] = E[X] = E[X\cdot 1] = \lim_{n\uparrow\infty} E[X_n\cdot 1] = \lim_{n\uparrow\infty} E[X_n] = 0.$$

It now remains to prove that

$$X_n \xrightarrow{\text{qm}} X, \qquad Y_n \xrightarrow{\text{qm}} Y \tag{47}$$

and this will be done using Cauchy's criterion for convergence in quadratic mean (Theorem T12).

The same computation as that leading to Eq. (46) yields

$$\lim_{n,m\uparrow\infty} E[X_n X_m] \equiv \int_0^1\int_0^1 f(u)f(v)\Gamma_X(u,v)\,du\,dv$$

and therefore, since

$$E[(X_n - X_m)^2] = E[X_n^2] + E[X_m^2] - 2E[X_n X_m],$$

we see that

$$\lim_{n, m \uparrow \infty} E[(X_n - X_m)^2] = 0.$$

Therefore, by Theorem T13, there exists a random variable $\tilde{X}$ such that $X_n \xrightarrow{\text{qm}} \tilde{X}$. One must now show that $\tilde{X} = X$, P-as, a not completely obvious fact. In view of the results of Section 3 summarized in the hierarchical tableau, from a sequence $(X_n, n \geqslant 1)$ such that $X_n \xrightarrow{\text{qm}} \tilde{X}$ one can extract a subsequence $(X_{n_h}, h \geqslant 1)$ such that $X_{n_h} \xrightarrow{\text{as}} \tilde{X}$. Since $X_{n_h} \xrightarrow{\text{as}} X$, we must have $X = \tilde{X}$, P-as, qed.

Filtered Gaussian Signals Are Gaussian. Recall that a random signal $[X(t), t \in \mathbb{R}]$ such that for all $t_1, \ldots, t_n \in \mathbb{R}$, the vector $[X(t_1), \ldots, X(t_n)]$ is Gaussian is called a Gaussian signal. Another way of stating this (Chapter 4, Definition D1) is to say that for all $t_1, \ldots, t_n \in \mathbb{R}$ and all $\lambda_1, \ldots, \lambda_n \in \mathbb{R}$, the random variable $\sum_{i=1}^n \lambda_i X(t_i)$ is Gaussian. Our purpose here is to demonstrate that if the input signal $[X(t), t \in \mathbb{R}]$ is Gaussian, the output signal defined by Eq. (39) is also Gaussian, i.e., that for every $t_1, \ldots, t_n \in \mathbb{R}$ and $\lambda_1, \ldots, \lambda_n \in \mathbb{R}$,

$$Z = \sum_{i=1}^n \lambda_i Y(t_i)$$

is Gaussian. This is actually a consequence of Theorem T15, as we shall see.

Recall that for any $t \in R$, $Y(t)$ is the limit in the quadratic mean of finite sums as in Eq. (44). Therefore, Z is the limit in the quadratic mean of finite linear combinations of the form $\sum_{j=1}^N \mu_j X(s_j)$. Since $[X(t), t \in \mathbb{R}]$ is a Gaussian signal, these combinations are Gaussian. Thus, Z is the limit in the quadratic mean of a sequence of Gaussian random variables, and therefore, by Theorem T15, Z is itself Gaussian qed.

SOLUTIONS FOR CHAPTER 5

E1. $\omega \in \bigcap_{n \geqslant 1} \bigcup_{k \geqslant n} A_k \Leftrightarrow \omega \in \bigcup_{k \geqslant n} A_k$ for all $n \geqslant 1 \Leftrightarrow$ for all $n \geqslant 1$, there exists $j \geqslant 1$ such that $\omega \in A_{n+j}$.

E2. Since the X_n's are independent, the direct and converse part of Borel–Cantelli's lemma reads

$$P(X_n = 1 \quad \text{io}) = 0 \Leftrightarrow \sum_{n \geqslant 1} P(X_n = 1) < \infty.$$

This proves the announced result. Indeed, take $N = \{X_n = 1 \quad \text{i.o.}\}$. We just saw that $P(N) = 0$ (because $\sum p_n < \infty$), and $(\omega \notin N) \Leftrightarrow [\text{there exists } K(\omega) \text{ such that } n > K(\omega) \Rightarrow X_n \neq 1]$. But $X_n \neq 1$ means $X_n = 0$.

E3.
$$P\left(\left|\frac{S_n}{n} - p\right| \geqslant \varepsilon\right) \leqslant \frac{1}{\varepsilon^4} E\left[\left(\frac{S_n - np}{n}\right)^4\right].$$

$$E\left[\left(\frac{S_n - np}{n}\right)^4\right] = \frac{1}{n^4} \sum_{i,j,k,l=1}^n E[(X_i - p)(X_j - p)(X_k - p)(X_l - p)].$$

Only the terms of the type $E[(X_i - p)^4]$ and $E[(X_i - p)^2(X_j - p)^2]$ remain in the latter sum. There are n terms of the type $E[(X_i - p)^4]$ and $E[(X_i - p)^4] = \alpha$ finite. There are $3n(n - 1)$ terms of the type $E[(X_i - p)^2(X_j - p)^2]$ $(i \neq j)$ and $E[(X_i - p)^2(X_j - p)^2] = \beta$ finite. Therefore,

$$E\left[\left(\frac{S_n - np}{n}\right)^4\right] \leqslant \frac{n\alpha + 3n(n - 1)\beta}{n^4}.$$

The right-hand side of the above inequality defines the general term of a convergent series ($\sim 3\beta/n^2$).

E4. According to the direct part of Borel–Cantelli's lemma, $P(|X_n - X| \geqslant \varepsilon_n$ i.o. $= 0$. Therefore, letting $N = \{|X_n - X| \geqslant \varepsilon_n$ i.o.$\}$, we have $P(N) = 0$ and if $\omega \notin N, |X_n(\omega) - X(\omega)| < \varepsilon_n$ for all n larger than some finite integer $N(\omega)$, which implies $X_n(\omega) \to X(\omega)$ since $\lim \varepsilon_n = 0$.

E5. Define $S_n = T_n - T_{n-1}$ $(n \geqslant 1)$. The sequence $(S_n, n \geqslant 1)$ is iid with common mean $1/\lambda$. Therefore, by the strong law of large numbers,

$$\frac{S_1 + \cdots + S_n}{n} \xrightarrow{\text{as}} 1/\lambda.$$

Since $S_1 + \cdots + S_n = T_n$ and $n = N(T_n)$, this means $N(T_n)/T_n \xrightarrow{\text{as}} \lambda$, and this implies $N(t)/t \xrightarrow{\text{as}} \lambda$ since $T_n \xrightarrow{\text{as}} \infty$ (because $T_n/n \xrightarrow{\text{as}} 1/\lambda > 0$).

E6. The sequence $(Y_{n(k+1)+i}, n \geqslant 1)$ is iid with $E[Y_{n(k+1)+i} = 1] = P(X_{n(k+1)+i} = x_0, \ldots, X_{n(k+1)+i+k} = x_k) = (\frac{1}{2})^{k+1}$, and therefore, by the strong law of large numbers,

$$\left(\sum_{n=1}^{N} Y_{n(k+1)+i}\right) \Big/ N \xrightarrow{\text{as}} \left(\frac{1}{2}\right)^{k+1}.$$

This being true for all i $(0 \leqslant i \leqslant k)$, it is not difficult to show that $(\sum_{k=1}^{n} Y_n)/n \xrightarrow{\text{as}} (\frac{1}{2})^{k+1}$. (Do the complete proof paying particular attention to the sets of null probability.)

E7. For $x \in [0, 1]$:

$$P(Z_n \leqslant x) = P(\min(X_1, \ldots, X_n) \leqslant x) = 1 - P(\min(X_1, \ldots, X_n) > x)$$

$$= 1 - P(X_1 > x, \ldots, X_n > x) = 1 - (1 - x)^n.$$

Therefore,

$$P(nZ_n \leqslant x) = P\left(Z_n \leqslant \frac{x}{n}\right) = 1 - \left(1 - \frac{x}{n}\right)^n \xrightarrow[n \uparrow \infty]{} 1 - e^{-x}.$$

E8. $\qquad P(X_1 + \cdots + X_n = k) = \binom{n}{k} p_n^k (1 - p_n)^{n-k}, (0 \leqslant k \leqslant n).$

Letting $P(X_1 + \cdots + X_n = k) = f(n, k)$, we see that $f(n, 0) = (1 - p_n)^n \xrightarrow[n \uparrow \infty]{} e^{-\lambda}$ since $\log(1 - p_n)^n = n\log(1 - p_n) \simeq -np_n \xrightarrow[n \uparrow \infty]{} -\lambda$. Also,

$$\frac{f(n, k + 1)}{f(n, k)} = \frac{p_n}{1 - p_n} \frac{k + 1}{n - k} = (k + 1)np_n \frac{1}{(1 - p_n)\left(1 - \frac{k}{n}\right)} \xrightarrow[n \uparrow \infty]{} \lambda(k + 1).$$

Therefore, by induction starting with $\lim_{n \uparrow \infty} f(n, 0) = e^{-\lambda}$, we obtain

$$\lim_{n \uparrow \infty} f(n, k) = e^{-\lambda} \frac{\lambda^k}{k!}, \ k \in \mathbb{N}.$$

E9.
$$\phi_n(u) = E[e^{iu\left(\sum_{i=1}^{n} x_i\right)/n^2}] = (Ee^{iu(X_1/n^2)})^n = (e^{|u|/n^2})^n,$$

i.e., $\phi_n(u) = e^{-|u|/n}$. Therefore,

$$\lim_{n \uparrow \infty} \phi_n(u) = \phi(u) = \begin{cases} 1 & \text{if } u = 0 \\ 0 & \text{if } u \neq 0 \end{cases}$$

and ϕ is *not continuous at* 0, therefore $(X_1 + \cdots + X_n)/n^2$ does *not* converge in law.

E10.
$$\phi_n(u) = E[e^{iuX_n}] = \frac{1}{n} \sum_{k=1}^{n} e^{iuk/n} = \frac{1}{n} e^{i(u/n)} \frac{1 - e^{iu}}{1 - e^{iu/n}}.$$

Therefore, since

$$\lim_{n \uparrow \infty} n(1 - e^{iu/n}) = -iu, \qquad \lim_{n \uparrow \infty} \phi_n(u) = \frac{e^{iu} - 1}{iu}.$$

Since $\phi(u) = (e^{iu} - 1)/iu$ is the cf of the uniform density on $[0, 1]$, $(X_n, n \geq 1)$ converges in law to the uniform distribution on $[0, 1]$.

E11.
$$\phi_n(u) = Ee^{iu\left\{\left(\sum_{i=1}^{n} X_i/\sqrt{npq}\right) - (np/\sqrt{npq})\right\}} = e^{+iu\sqrt{n(p/q)}} E[e^{iu(X_1/\sqrt{npq})}]^n$$

$$= e^{iu\sqrt{n(p/q)}}(q + pe^{iu(u/\sqrt{npq})})^n = e^{iu\sqrt{n(p/q)}}[1 + p(e^{iu/\sqrt{npq}} - 1)]^n.$$

We do the case $p = q = \frac{1}{2}$ for simplicity:

$$\phi_n(u) = e^{iu\sqrt{n}}[1 + \frac{1}{2}(e^{2iu/\sqrt{n}} - 1)]^n$$

$$\log \phi_n(u) = iu\sqrt{n} + \{n \log[1 + \frac{1}{2}(e^{2iu/\sqrt{n}} - 1)]\} = -\frac{1}{2}u^2[1 + \varepsilon(n)]$$

with $\lim_{n \uparrow \infty} \varepsilon(n) = 0$. Therefore, $\phi_n(u) \to_{k \uparrow \infty} e^{-1/2u^2}$ which is the cf of $\mathcal{N}(0, 1)$.

E12. Let $(X_n, n \geq 1)$ be a sequence of iid Poisson random variables with mean 1 and variance 1. Then $X_1 + \cdots + X_n$ is Poisson, mean n, and therefore

$$e^{-n} \sum_{k=0}^{n} \frac{n^k}{k!} = P(X_1 + \cdots + X_n \geq n) = P\left(\frac{X_1 + \cdots + X_n - n}{\sqrt{n}} \geq 0\right).$$

By the central limit theorem this quantity tends to $(1/\sqrt{2\pi}) \int_0^\infty e^{-1/2y^2} \, dy = \frac{1}{2}$.

E13. Let $(X_n, n \geq 1)$ and $(Y_n, n \geq 1)$ be two sequences of random variables with the same cdf F, and such that the family $(X_n, Y_n, n \geq 1)$ is a family of independent random variables. Define for each $n \geq 1$

$$Z_n = \frac{X_1 + \cdots + X_{2n-1} + Y_1 + \cdots + Y_{2n-1}}{\sqrt{2^n}}.$$

By induction, $P(Z_n \leq x) = F(x)$. By the central limit theorem (written with 2^n instead of n), $Z_n \xrightarrow{\mathscr{L}} \mathcal{N}(0, 1)$, hence the announced result.

E14. We do the first inequality only. Since $X \leqslant x - \varepsilon$ and $|X_n - X| < \varepsilon$ imply $X_n \leqslant x$,

$$\{X \leqslant x - \varepsilon\} - \{X_n \leqslant x - \varepsilon\} \cap \{|X_n - X| \geqslant \varepsilon\} \subset \{X_n \leqslant x\},$$

and therefore,

$$P(X \leqslant x - \varepsilon) - P(X_n \leqslant x - \varepsilon, |X_n - X| \geqslant \varepsilon) \leqslant P(X \leqslant x).$$

But from Term $(*)$,

$$P(X_n \leqslant x - \varepsilon, |X_n - X| \geqslant \varepsilon) \leqslant P(|X_n - X| \geqslant \varepsilon) \leqslant \delta.$$

E15. Since $X_n \xrightarrow{\text{as}} X$, $|X_n - X|^2 \xrightarrow{\text{as}} 0$. If $|X_n| \leqslant Y$, P-as for all $n \geqslant 1$, also $|X| \leqslant Y$, P-as, and therefore $|X_n - X|^2 \leqslant 4Y^2$, P-as. The rest follows by Lebesgue's dominated convergence theorem.

E16.
$$E[X_m Y_n] - E[XY] = E[(X_m - X)Y] + E[X_m(Y_n - Y)].$$

From Schwarz's inequality

$$|E[(X_m - X)Y]|^2 \leqslant E[Y^2] \cdot E[(X_m - X)^2] \to 0 \qquad \text{as} \quad m \to \infty$$

$$|E[X_m(Y_n - Y)]|^2 \leqslant E[X_m^2] \cdot E(Y_n - Y)^2] \to 0 \qquad \text{as} \quad m, n \to \infty$$

if $E[X_m^2]$ is bounded in m. This is the case because

$$E[X_m^2] \leqslant 2E[(X_m - X)^2] + 2E(X^2)$$

and $\lim_{m \uparrow \infty} E[(X_m - X)^2] = 0$. Therefore, $\lim_{m,n \uparrow \infty} E[X_m Y_n] = E[XY]$, qed. Applying the result to X_n and $Y_n = X_n$ yields $\lim_{n \uparrow \infty} E[X_n^2] = E(X^2)$. If $Y_n = 1$ (which converges in all senses to 1!), we obtain

$$\lim_{n \uparrow \infty} E[X_n 1] = E[X1],$$

i.e.,

$$\lim_{n \uparrow \infty} E[X_n] = E[X].$$

E17. For all $\varepsilon > 0$,

$$\lim_{n \uparrow x} P(|X_n - c| \geqslant \varepsilon) = \lim_{n \uparrow \infty} P(X_n \geqslant c + \varepsilon) + P(X_n \leqslant c - \varepsilon)\}$$

$$= 1 + \lim_{n \uparrow x} P(X_n \leqslant c - \varepsilon) - P(X_n < c - \varepsilon)\}$$

$$= 1 - \{0 - 1\} = 0.$$

Additional Exercises

Note: Exercises 1 to 16 are relative to Chapters 1, 2, and 3. Chapters 4 and 5 are needed for Exercises 17 to 28.

Exercise 1 (A Sequence of Liars). Consider a sequence of n "liars" $L_1, \ldots, L_n$. The first liar L_1 receives information about the occurrence of some event ("yes" or "no") and transmits it to L_2, who transmits it to L_3, etc .... Each liar transmits what he hears with probability $p(0 < p < 1)$, and the contrary with probability $q = 1 - p$. The decision of lying or not lying is made independently by each liar. What is the probability p_n to obtain the correct information from L_n? What happens when n increases to infinity?

Exercise 2 (The Golden Ring). There are $2n$ bits of thread:

Two people, operating independently of each other, make knots. The first one makes knots on the upper extremities, and the other one on the lower extremities. Each lower (resp. upper) extremity is involved in one and only one knot. For instance with $2n = 6$, you can obtain, among other configurations:

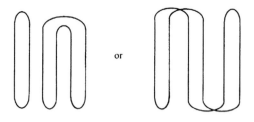

In the second situation, you have just one piece of thread forming a ring. What is the probability P_{2n} for this event to occur in the general case?

Hint: Find P_2 and a recurrence relation between P_{2n} and P_{2n-2}.

Exercise 3 (Shall I Get This Book?). You are looking for a book in the campus libraries. Each library has it with probability 0.60, but the book may have been borrowed by some other patron with probability 0.25. If there are 3 libraries, what are your chances of obtaining the book?

Exercise 4 (Winning a Game of Heads and Tails). Two players A and B with respective initial fortunes $\$a$ and $\$b$ (a, b strictly positive integers) play a game of heads and tails, betting an amount of \$1 at each toss. The outcome "heads" has probability $p(0 < p < 1)$. Player A wins on "heads." Compute the probability for A to win (the game ends when one of the players is broke).

Hint: You must compute the probability $u(a)$ that A starting with a fortune of $x = a$ reaches fortune $c = a + b$ without getting broke in the meantime. Of course $x = a$ is of interest, but you will compute the probability $u(x)$ for all integers $x \in [0, c]$. To do this derive a recurrence relation for $u(x)$, and solve it.

Exercise 5 (Heads and Tails Again). A person, named A, throws an *unbiased* coin N times and obtains T_A "tails." Another person, B, throws his own unbiased coin $N + 1$ times and has T_B "tails." What is the probability that $T_A \geq T_B$?

Hint: Introduce H_A and H_B the number of "heads" obtained by A and B respectively, and use a symmetry argument.

Exercise 6 (Boys and Girls Are Independent). Let X be a Poisson random variable with mean $\lambda > 0$, independent of $(Y_n, n \geq 0)$, a sequence of $\{0, 1\}$ valued iid random variables with $P(Y_n = 1) = p$. Show that $U = \sum_{n=1}^{X} Y_n$ and $V = X - U$ are independent Poisson random variables of mean λp and $\lambda(1 - p)$ respectively.

Remark 1. This is generally considered a "paradoxical" result since it would be obviously false if X were replaced by a fixed number k.

Remark 2. The title refers to a model where the total progeny of a couple consists of X boys and girls, with U boys and V girls.

Exercise 7 (Group Testing). N individuals must undergo a medical test which is somewhat expensive. Instead of analyzing a blood sample for each of them (a procedure which would require N separate analyses), the blood samples are mixed and the test is performed on this "group sample." If it is positive (i.e., if at least one of the patients has a positive reaction), then all N patients undergo an individual test. The probability of a positive reaction for a given patient is p, and it is assumed that, with respect to this test, the patients react independently of one another. What is the average number of tests required in the group testing procedure? Compare it to N when $N = 4$, $p = 1/10$.

Exercise 8 (Decoding a Characteristic Function). Find the distribution of a random variable X with cf $\phi_X(u) = 1/(2e^{-iu} - 1)$.

Exercise 9 (The Covariance Matrix of a Multinomial Vector). Let $X = (X_1, \ldots, X_k)$ be a k-dimensional random vector where the X_i's take their values in $\mathbb{N}$. Let $g(s_1, \ldots, s_k)$ be the generating function of this vector. Show that $E[X_i X_j - \delta_{ij}] = (\partial^2 g/\partial s_i \partial s_j)(1, \ldots, 1)$. Apply this to vector $X \sim \mathcal{M}(n, k, p_i)$ (Eq. (11) of Chapter 2), for which it is known (E21 of Chapter 2) that $g(s_1, \ldots, s_k) = (p_1 s_1 + \cdots + p_k s_k)^n$, to compute its covariance matrix.

Exercise 10 (The Random Pen Club). You write n personal letters to be sent to n of your friends and you write the addresses at random on the envelopes. What is the probability that at least one of the envelopes has the correct adress? What if n is very large?

Exercise 11 (The Matchbox (Banach's Problem)). A smoker has one matchbox with n matches in each pocket. He reaches at random for one box or the other. What is the probability that, having eventually found an empty matchbox, there will be k matches left in the other box?

Exercise 12 (Infinite Expectation in the Coin Tossing Game). Two players toss an unbiased coin in turn until both get the same number of heads. Let $2N$ be the total number of coin tosses needed for equalization. Find $P(2N = 2n)$ and give the expectation of $2N$.

Exercise 13 (Lottery Tickets). Lottery tickets have numbers going from $0\,0\,0\,0\,0\,0$ to $9\,9\,9\,9\,9\,9$. Find the probability of purchasing a ticket with the sum of the first three digits equal to the sum of the last three digits.
Hint: Use generating functions.

Exercise 14 (The Uniform Distribution on a Disk). The following model of a point selected at random uniformly in the closed unit disk D of R^2 centered at the origin is proposed: Take $\Omega = D$ and let $\mathcal{F}$ be the family of subsets of D for which the area can be defined, and for each set of A of $\mathcal{F}$ with area

$S(A)$ define $P(A) = S(A)/S(\Omega) = S(A)/\pi$. The figure below introduces a few notations.

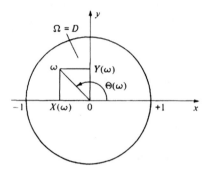

Define $Z = (X^2 + Y^2)^{1/2}$. Thus Z is a random variable taking its values in $[0,1]$ and Θ takes its values in $[0,2\pi)$. Compute $P(0 \leqslant a \leqslant Z \leqslant b \leqslant 1, 0 \leqslant \theta_1 \leqslant \Theta < \theta_2 < 2\pi)$ and show that Z and Θ are independent.

Exercise 15 (Equation with Random Coefficients). The numbers a and b are selected independently and uniformly on the segment $[-1, +1]$. Find the probability that the roots of the equation $x^2 + 2ax + b$ are real.

Exercise 16 (Infinum of Exponential Random Variables). Let $X_1, \ldots, X_n$ be independent exponential random variables, with $E[X_i] = 1/\lambda_i$ $(1 \leqslant i \leqslant n)$. Find the pdf of $Y = \inf(X_1, \ldots, X_n)$.

Exercise 17 (Gaussian Vector). Let X and Y be two independent standard Gaussian random variables. Show that the two-dimensional random vector $((X + Y)/\sqrt{2}, (X - Y)/\sqrt{2})$ is a standard Gaussian vector, *without using the formula of change of variables*.

Exercise 18 (Spherical Coordinates). The random variables $X_1, \ldots, X_n$ are iid $\mathcal{N}(0, \sigma^2)$. Find the pdf of the vector $R, \Phi_1, \ldots, \Phi_{n-1}$ where $R \geqslant 0$, $\Phi_i \in [0, 2\pi)$ $(1 \leqslant i \leqslant n - 1)$, and

$$
\begin{aligned}
X_1 &= R \sin \Phi_1 \\
X_2 &= R \sin \Phi_2 && \cos \Phi_1 \\
X_3 &= R \sin \Phi_3 && \cos \Phi_2 && \cos \Phi_1 \\
&\vdots \\
X_{n-1} &= R \sin \Phi_{n-1} && \cos \Phi_{n-2} && \ldots\ldots && \cos \Phi_1 \\
X_n &= R \cos \Phi_{n-1} && \cos \Phi_{n-2} && \ldots\quad\ldots && \cos \Phi_1.
\end{aligned}
$$

Exercise 19 (Integral Powers of Cauchy Random Variables). Let X be a random variable with the pdf $(1/\pi)(a/a^2 + x^2)$. Find the pdf of $Y = X^n$ where n is a positive integer.

Exercise 20 (Change of Variables with Exponential Random Variables). Let $U_1, \ldots, U_{n+1}$ be $n + 1$ iid random variables with pdf $f(u) = \lambda e^{-\lambda u}$ if $u > 0$, $f(u) = 0$ if $u \leqslant 0$. Define $Y_i = U_i/(U_1 + \cdots + U_{n+1})$ for $1 \leqslant i \leqslant n$ and $S_n = U_1 + \cdots + U_{n+1}$. Compute the pdf's of the vectors $(Y_1, \ldots, Y_n, S_n)$ and $(Y_1, \ldots, Y_n)$, and of the random variable nY_i.

Exercise 21 (Poisson Process and Binomial Random Variable). Let $(N(t), t \geqslant 0)$ be the counting process of a homogeneous Poisson process $(T_n, n \geqslant 1)$ with intensity $\lambda > 0$. What is the conditional distribution of $N(s)$ given $N(t) = n$ $(s \leqslant t)$? [i.e., compute $P(N(s) = k | N(t) = n)$].

Exercise 22 (The Flip-Flop Stochastic Process). Let $(N(t), t \geqslant 0)$ be the counting process of a homogeneous Poisson process $(T_n, n \geqslant 1)$ with intensity $\lambda > 0$. Define for each t the random variable $X(t)$ by $X(t) = X(0)(-1)^{N(t)}$ where $X(0)$ is a random variable taking the values -1 and $+1$, independent of $(N(t), t \geqslant 0)$. Therefore for each $\omega \in \Omega$, $t \to X(t, \omega)$ is a function flipping from -1 to $+1$ and flopping from $+1$ to -1:

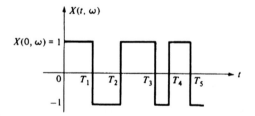

When $P(X(0) = 1) = p$, compute $P(X(t) = 1)$ for each $t \geqslant 0$. Find $\lim_{t \uparrow \infty} P(X(t) = 1)$.

Exercise 23 (A Poisson Series). Use the central limit theorem to prove that

$$\lim_{n \to \infty} \sum_{k=0}^{n} e^{-n} \frac{n^k}{k!} = \frac{1}{2}.$$

Exercise 24 (Standardized Poisson Counting Process). Let $(N_t, t \geqslant 0)$ be the counting process of a homogeneous Poisson process with intensity $\lambda > 0$. Compute the cf of $(N_t - \lambda t)/\sqrt{\lambda t}$ and show that the latter rv converges in distribution to $\mathcal{N}(0, 1)$ as t goes to ∞.

Exercise 25 (Asymptotic Estimate of the Intensity of a Homogeneous Poisson Process). Use the strong law of large numbers to show that if $(N_t, t \geqslant 0)$ is the counting process of a homogeneous Poisson process with intensity $\lambda > 0$, then $\lim_{t \to \infty} N_t = \infty$, P — as. Show that $\lim_{t \to \infty} N_t/t = \lambda$, P — as.

Exercise 26 (Feller's Paradox). Let $(T_n, n \geqslant 1)$ be a homogeneous Poisson process over $\mathbb{R}_+$, with intensity $\lambda > 0$. Let $T_0 = 0$. For a *fixed* $t > 0$, define $U_t = t - T_{N_t}$, $V_t = T_{N_{t-1}} - t$ (see figure below).

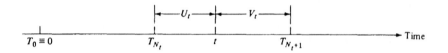

Note that T_{N_t} is the first random point T_n strictly to the left of t and $T_{N_{t+1}}$ is the first random point T_n after t. Find the joint distribution of U_t and V_t. Find the distribution of $T_{N_{t+1}} - T_{N_t}$. Examine the case $t \to \infty$.

Remark. Feller's paradox consists in observing that although the random variables $T_{n+1} - T_n (n \geqslant 0)$ have the same exponential distribution, $T_{N_{t+1}} - T_{N_t}$ is not an exponential random variable.

Exercise 27 (Sum of Sums of Gaussian Random Variables). Let $(\varepsilon_n, n \geqslant 1)$ be a sequence of iid standard Gaussian random variables and define the sequence $(X_n, n \geqslant 1)$ by: $X_n = \varepsilon_1 + \cdots + \varepsilon_n$. Show that $(X_1 + \cdots + X_n)/n\sqrt{n} \xrightarrow{\mathscr{L}} N(0, \sigma^2)$ for some σ to be computed.

Exercise 28 (Gaussian Random Variables Converging in Law). Let $(U_n, n \geqslant 0)$ be a sequence of iid random variables, Gaussian, mean 0 and variance 1. Define $(X_n, n \geqslant 0)$ by $X_0 = U_0$, $X_{n+1} = aX_n + U_{n+1}$ $(n \geqslant 0)$.

(i) Show that X_n is a Gaussian random variable, and that if $a < 1$, $X_n \xrightarrow{\mathscr{L}} \mathscr{N}(0, \sigma^2)$ for some σ^2 to be identified.

(ii) If $a > 1$ show that X_n/a^n converges in quadratic mean. Does it converge in distribution?

Solutions to Additional Exercises

1. $p_n = (1 - (p - q)^n)/2$, $\lim_{n \uparrow \infty} p_n = \frac{1}{2}$.

2. $P_2 = 1$, $P_{2n} = P_{2n-2}(2n - 2)/(2n - 1)$. To see this, consider thread 1 and its "upper associate"

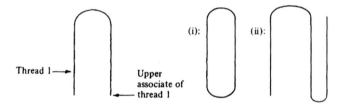

(i):   (ii):

Thread 1 ⟶

Upper
associate of
⟵ thread 1

Then either (i) or (ii) occurs. (ii) occurs with probability $(2n - 2)/(2n - 1)$. In situation (ii), the piece has become one bit of thread, to be linked to the $2n - 3$ remaining pieces. Finally $P_{2n} = 2 \cdot 4 \cdot 6 \ldots (2n - 2)/1 \cdot 3 \cdot 5 \ldots (2n - 1)$.

3. $1 - [(0.60)(0.25) + 0.40]^3 = 1 - (0.55)^3$.

4. If $x = 0$, then clearly $u(0) = 0$. If $x = c$, then clearly $u(c) = 1$. For $0 < x < c$, after a toss of the coin, player A has fortune $x + 1$ with probability p, and fortune $x - 1$ with probability q. From $x + 1$ (resp. $x - 1$), the probability of reaching c without getting broke in the meantime is $u(x + 1)$ (resp. $u(x - 1)$), and therefore (exclusive and exhaustive causes)

$$u(x) = pu(x + 1) + qu(x - 1).$$

The solution of this recurrence equation with boundary conditions $u(0) = 0$, $u(c) = 1$ is (computations) for $x = a$:

$$u(a) = \begin{cases} a/(a+b) & \text{if } p = q = 1/2 \\ [1 - (q/p)^a]/[1 - (q/p)^{a+b}] & \text{if } p \neq q. \end{cases}$$

5. By symmetry (since the coins are *unbiased*), $P(T_A \geq T_B) = P(H_A \geq H_B)$. But $H_A = N - T_A$ and $H_B = N + 1 - T_B$ so that $H_A \geq H_B \Leftrightarrow T_A < T_B$. Therefore $P(H_A \geq H_B) = P(T_A < T_B) = 1 - P(T_A \geq T_B)$ and, using the first equality, $P(T_A \geq T_B) = 1 - P(T_A \geq T_B)$, i.e., $P(T_A \geq T_B) = 1/2$.

6. $P(U = k, V = l) = P(U = k, X = k + l) = P(X = k + l, Y_1 + \cdots + Y_{k+l} = k)$

$$= P(X = k + l)P(Y_1 + \cdots + Y_{k+l} = k)$$

$$= e^{-\lambda}\frac{\lambda^{k+l}}{(k+l)!} \times \frac{(k+l)!}{k!l!}p^k q^l$$

$$= e^{-\lambda p}\frac{(\lambda p)^k}{k!} \times e^{-\lambda q}\frac{(\lambda q)^l}{l!}.$$

7. $1 + (1 - (1 - p)^N)N$. Numerical application: 2.3756.

8. $P(X = k) = 2^{-k}$ (k integer ≥ 1).

9. We do the case $i = 1, j = 2$. The generating function of (X_1, X_2) is $f(s_1, s_2) = g(s_1, s_2, 1, \ldots, 1)$. Also $f(s_1, s_2) = \sum_{k_1 \in \mathbb{N}} \sum_{k_2 \in \mathbb{N}} P(X_1 = k_1, X_2 = k_2)s_1^{k_1}s_2^{k_2}$, so that $(\partial^2 f/\partial s_1 \partial s_2)(1, 1) = \sum_{k_1 \in \mathbb{N}} \sum_{k_2 \in \mathbb{N}} k_1 k_2 P(X_1 = k_1, X_2 = k_2) = E[X_1 X_2]$. The multinomial case: $E[X_i X_j] = n(n-1)p_i p_j$ if $i \neq j$ and $\text{Var}(X_i^2) = np_i(1 - p_i)$ (Why?). Also $E[X_i] = np_i$. Therefore $\Gamma_x = \{\gamma_{ij}\}$ with $\gamma_{ij} = -np_i p_j$ if $i \neq j$, $\gamma_{ii} = np_i(1 - p_i)$.

10. $A_k = (k^{\text{th}}$ envelope has the correct address). To be computed $P(\bigcup_{k=1}^n A_k)$. But

$$P(A_k) = \frac{1}{n} = \frac{(n-1)!}{n!}, \quad P(A_k \cap A_j) = P(A_k)P(A_j|A_k) = \frac{(n-2)!}{n!},$$

$$P(A_k \cap A_j \cap A_i) = \frac{(n-3)!}{n!}, \ldots, P\left(\bigcap_{k=1}^n A_k\right) = \frac{1}{n!}$$

Therefore the result is (inclusion–exclusion formula)

$$\binom{n}{1}\frac{(n-1)!}{n!} - \binom{n}{2}\frac{(n-2)!}{n!} + \binom{n}{3}\frac{(n-3)!}{n!} - \cdots + (-1)^{n-1}\frac{1}{n!}.$$

That is to say

$$1 - \frac{1}{2!} + \frac{1}{3!} - \cdots + (-1)^{n-1}\frac{1}{n!}.$$

If n is very large, this is close to $1 - e^{-1}$.

11. $\binom{2n-k}{n} / 2^{2n-k}$.

12. $P(2N = 2n) = \frac{(2n-2)!}{n!(n-1)!} \cdot \frac{1}{2^{2n-1}}; E[2N] = +\infty$

(use the equality $\sum_{n=0}^{\infty} \left(\frac{x}{4}\right)^n \frac{(2n)!}{(n!)^2} = (1 - x)^{-1/2}$).

13. Call $X_1 X_2 X_3 X_4 X_5 X_6$ the random number. The X_i's are independent and identically distributed random variables, with $P(X_i = k) = p = 1/10$ for $k = 0, \ldots, 9$. The generating functions of the X_i's are identical, equal to

$$\frac{1}{10}(1 + s + \cdots + s^9) = \frac{1}{10} \cdot \frac{1 - s^{10}}{1 - s}.$$

The generating function $g(s)$ of $X_1 + X_2 + X_3$ is

$$\frac{1}{10^3} \cdot \frac{(1 - s^{10})^3}{(1 - s)^3}$$

and is equal to the gf of $X_4 + X_5 + X_6$. The coefficient of s^r in $g(s)$ is the probability that the sum of the first 3 digits is r, and the coefficient of s^{-r} in $g(s^{-1})$ is the probability that the sum of the last 3 digits is r. Therefore the coefficient of s^0 in $g(s)g(s^{-1})$ is the probability that we are looking for. We have

$$g(s)g(s^{-1}) = \frac{1}{10^6} \cdot \frac{1}{s^{27}} \left(\frac{1 - s^{10}}{1 - s} \right)^6.$$

But

$$(1 - s^{10})^6 = 1 - \binom{6}{1} s^{10} + \binom{6}{2} s^{20} + \cdots$$

and

$$(1 - s)^{-6} = 1 + \binom{6}{5} s + \binom{7}{5} s^2 + \cdots$$

and therefore the answer is

$$\frac{1}{10^6} \left(\binom{32}{5} - \binom{6}{1}\binom{22}{5} + \binom{6}{2}\binom{12}{5} \right) = 0.05525.$$

14. $P(a \leqslant Z \leqslant b, \theta_1 \leqslant \Theta < \theta_2) = (b^2 - a^2)(\theta_2 - \theta_1)/2\pi$. Therefore $P(a \leqslant Z \leqslant b) = b^2 - a^2$ (take $\theta_1 = 0, \theta_2 = 2\pi$), and $P(\theta_1 \leqslant \Theta \leqslant \theta_2) = (\theta_2 - \theta_1)/2\pi$ (take $a = 0, b = 1$).

15. $P(a^2 - b \geqslant 0) = \frac{1}{2} + \frac{1}{4} \cdot 2 \int_0^1 x^2 \, dx = \frac{2}{3}$.

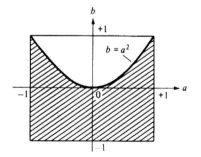

16. If $y \geqslant 0$, $1 - P(Y > y) = 1 - P(X_1 > y, \ldots, X_n > y) = 1 - e^{-\lambda_1 y} \ldots e^{-\lambda_n y}$. Therefore $f_Y(y) = (\lambda_1 + \cdots + \lambda_n)e^{-(\lambda_1 + \cdots + \lambda_n)y}$ ($y \geqslant 0$).

17. $((X + Y)/\sqrt{2}, (X - Y)/\sqrt{2})$ is Gaussian, because it is a linear function of the Gaussian vector (X, Y). (By the way, why is (X, Y) Gaussian?) $(X + Y)/\sqrt{2}$ and $(X - Y)/\sqrt{2}$ are independent because they are uncorrelated, and their variance is 1(compute).

18. $f(r, \phi_1, \ldots, \phi_{n-1}) = \dfrac{r^{n-1} \cos^{n-2} \phi_1 \cos^{n-3} \phi_2 \cdots \cos \phi_{n-2}}{(2\pi)^{n/2} \sigma^n} e^{-r^2/2\sigma^2}.$

19. If n is odd: $f_Y(y) = \dfrac{ay^{-(n-1)/n}}{\pi n(a^2 + y^{2/n})}.$

   If n is even: $f_Y(y) = \begin{cases} \dfrac{2ay^{-(n-1)/n}}{\pi n(a^2 + y^{2/n})} & \text{if } y \geqslant 0 \\ 0 & \text{otherwise.} \end{cases}$

20. $f_{Y_1, \ldots, Y_n, S_n}(y_1, \ldots, y_n, s)$

   $= \begin{cases} \lambda^{n+1} s^n e^{-\lambda s} & \text{if } (y_1, \ldots, y_n, s) \in \mathbb{R}_+^{n+1} \text{ and } 0 \leqslant y_1 + \cdots + y_n \leqslant 1 \\ 0 & \text{otherwise} \end{cases}$

   $f_{Y_1, \ldots, Y_n}(y_1, \ldots, y_n) = \begin{cases} n! & \text{if } (y_1, \ldots, y_n) \in \mathbb{R}_+^n \text{ and } 0 \leqslant y_1 + \cdots + y_n \leqslant 1 \\ 0 & \text{otherwise} \end{cases}$

   $f_{nY_i}(z) = \begin{cases} (1 - z/n)^{n-1} & \text{if } 0 \leqslant z \leqslant n \\ 0 & \text{otherwise.} \end{cases}$

21. $(s/t)^k (1 - s/t)^{n-k} n!/(k!(n-k)!)$ $(k = 0, \ldots, n).$

22. $1/2 + (p - 1/2)e^{-2\lambda t} \rightarrow 1/2.$

23. If $(X_n, n \geqslant 1)$ are iid Poisson random variables with mean 1 and variance 1,

   $$\lim_{n \uparrow \infty} P\left(\frac{(X_1 + \cdots + X_n) - n}{\sqrt{n}} \geqslant 0\right) = \frac{1}{2} \qquad \text{(central limit theorem)}.$$

   Observe that the sum $X_1 + \cdots + X_n$ is Poisson with mean n, and the conclusion follows.

24. The cf of $(N_t - \lambda t)/\sqrt{\lambda t}$ is $\exp\{\lambda t(e^{iu/\sqrt{\lambda t}} - 1 - iu/\sqrt{\lambda t})\} \rightarrow \exp\{-\frac{1}{2}u^2\}$ as $t \rightarrow +\infty.$

25. From the law of large numbers:

   $$\frac{N_n}{n} = \frac{1}{n} \sum_{k=0}^{n-1} [N_{k+1} - N_k] \xrightarrow[n \to \infty]{as} E[N_1] = \lambda > 0,$$

   therefore

   $$\lim_{t \to \infty} N_t = \lim_{n \to \infty} N_n = +\infty, \qquad P - \text{as.}$$

   Observe that

   $$\frac{N_t}{S_1 + \cdots + S_{N_t} + S_{N_t+1}} \leqslant \frac{N_t}{t} \leqslant \frac{N_t}{S_1 + \cdots + S_{N_t}}.$$

Also

$$\lim_{t\to\alpha}\frac{S_1 + \cdots + S_{N_t}}{N_t} = \lim_{n\to\alpha}\frac{S_1 + \cdots + S_n}{n}$$

since $N_t \to \infty$ as. The latter limit is $E[S_1] = 1/\lambda$ (law of large numbers). The conclusion follows from the above remarks.

26. $\{U_t \leq u, V_t \leq v\} = \{N_t - N_{t-u} \geq 1, N_{t+v} - N_t \geq 1\}$ $(0 \leq u < t, v \geq 0)$.

Therefore, for $0 \leq u < t, v \geq 0$,

$$P(U_t \leq u, V_t \leq v) = P(N_t - N_{t-u} \geq 1)P(N_{t+v} - N_t \geq 1) = (1 - e^{-\lambda u})(1 - e^{-\lambda v}).$$

Also for $u = t, v \geq 0$:

$$P(U_t = t, V_t \leq v) = P(N_t = 0, N_{t+v} - N_t \geq 1) = e^{-\lambda t}(1 - e^{-\lambda v}).$$

Finally for $u \geq 0, v \geq 0$:

$$P(U_t \leq u, V_t \leq v) = (1_{\{u \geq t\}} + 1_{\{u < t\}}(1 - e^{-\lambda u}))(1 - e^{-\lambda v}).$$

Therefore U_t and V_t are independent, V_t is an exponential rv of mean $1/\lambda$, whereas U_t as the cdf given by the figure below.

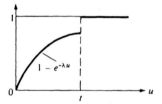

If $t \to \infty$, U_t converges in law to an exponential rv with mean $1/\lambda$. In the limit $t \to \infty$, $T_{N_{t+1}} - T_{N_t}$ is the sum of two independent exponential rv's with mean $1/\lambda$.

27. The cf of the vector $(X_1, \ldots, X_n)$ is $\phi_{X_1 \ldots X_n}(u_1, \ldots, u_n) = \exp\{-\frac{1}{2}[u_n^2 + (u_n + u_{n-1})^2 + \cdots + (u_1 + \cdots + u_n)^2]\}$. The cf of $(X_1 + \cdots + X_n)/n\sqrt{n}$ is $\exp\{-\frac{1}{2}u^2[1/n^3(1^2 + 2^2 + \cdots + n^2)]\}$ and tends to $\exp\{-\frac{1}{2}\cdot(u^2/3)\}$ as n goes to ∞; $\sigma^2 = 1/3$.

28. (i) X_n is a linear combination of $U_0, \ldots, U_n$ and $(U_0, \ldots, U_n)$ is a Gaussian vector since $U_0, \ldots, U_n$ are independent Gaussian vectors. Therefore X_n is Gaussian. $EX_n = 0$ and therefore

$$\phi_{X_n}(u) = e^{-(1/2)/E[X_n^2]u^2}.$$

But $E[X_{n+1}^2] = a^2 E[X_n^2] + E[U_{n+1}^2] = a^2 E[X_n^2] + 1$. Therefore since $E[X_0^2] = E[U_0^2] = 1$, $\lim_{n\to\infty} E[X_n^2] = 1/(1 - a^2)$, and

$$\lim \phi_{X_n}(u) = e^{-(1/2)/(1/1 - a^2)u^2} \qquad \text{i.e., } X_n \xrightarrow{\mathscr{L}} N\left(0, \frac{1}{1 - a^2}\right).$$

(ii) $\dfrac{X_{n+1}}{a^{n+1}} = \dfrac{X_n}{a^n} + \dfrac{U_{n+1}}{a^{n+1}}.$

Therefore

$$\frac{X_n}{a_n} = U_0 + \frac{U_1}{a} + \cdots + \frac{U_n}{a^n}.$$

Letting

$$Z_n = \frac{X_n}{a^n},$$

we have

$$E[|Z_{m+n} - Z_n|^2] = E\left[\left|\frac{U_{n+1}}{a^{n+1}} + \cdots + \frac{U_{n+m}}{a^{n+m}}\right|\right]$$

$$= \frac{1}{a^{n+1}}\left(1 + \cdots + \frac{1}{a^{m-1}}\right) = \frac{1}{a^{n+1}} \frac{1 - \frac{1}{a^m}}{1 - \frac{1}{a}} \to 0 \text{ as } n, m \to \infty.$$

Thus $(Z_n, n \geq 0)$ converges in quadratic mean (Theorem T13). Convergence in qm implies convergence in law.

Index

Undergraduate Texts in Mathematics

(continued from page ii)

Kemeny/Snell: Finite Markov Chains.

Kinsey: Topology of Surfaces.

Klambauer: Aspects of Calculus.

Lang: A First Course in Calculus. Fifth edition.

Lang: Calculus of Several Variables. Third edition.

Lang: Introduction to Linear Algebra. Second edition.

Lang: Linear Algebra. Third edition.

Lang: Undergraduate Algebra. Second edition.

Lang: Undergraduate Analysis.

Lax/Burstein/Lax: Calculus with Applications and Computing. Volume 1.

LeCuyer: College Mathematics with APL.

Lidl/Pilz: Applied Abstract Algebra.

Macki-Strauss: Introduction to Optimal Control Theory.

Malitz: Introduction to Mathematical Logic.

Marsden/Weinstein: Calculus I, II, III. Second edition.

Martin: The Foundations of Geometry and the Non-Euclidean Plane.

Martin: Transformation Geometry: An Introduction to Symmetry.

Millman/Parker: Geometry: A Metric Approach with Models. Second edition.

Moschovakis: Notes on Set Theory.

Owen: A First Course in the Mathematical Foundations of Thermodynamics.

Palka: An Introduction to Complex Function Theory.

Pedrick: A First Course in Analysis.

Peressini/Sullivan/Uhl: The Mathematics of Nonlinear Programming.

Prenowitz/Jantosciak: Join Geometries.

Priestley: Calculus: An Historical Approach.

Protter/Morrey: A First Course in Real Analysis. Second edition.

Protter/Morrey: Intermediate Calculus. Second edition.

Roman: An Introduction to Coding and Information Theory.

Ross: Elementary Analysis: The Theory of Calculus.

Samuel: Projective Geometry. *Readings in Mathematics.*

Scharlau/Opolka: From Fermat to Minkowski.

Sethuraman: Rings, Fields, and Vector Spaces: An Approach to Geometric Constructability.

Sigler: Algebra.

Silverman/Tate: Rational Points on Elliptic Curves.

Simmonds: A Brief on Tensor Analysis. Second edition.

Singer/Thorpe: Lecture Notes on Elementary Topology and Geometry.

Smith: Linear Algebra. Second edition.

Smith: Primer of Modern Analysis. Second edition.

Stanton/White: Constructive Combinatorics.

Stillwell: Elements of Algebra: Geometry, Numbers, Equations.

Stillwell: Mathematics and Its History.

Stillwell: Numbers and Geometry. *Readings in Mathematics.*

Strayer: Linear Programming and Its Applications.

Thorpe: Elementary Topics in Differential Geometry.

Toth: Glimpses of Algebra and Geometry.

Troutman: Variational Calculus and Optimal Control. Second edition.

Valenza: Linear Algebra: An Introduction to Abstract Mathematic:

Whyburn/Duda: Dynamic Topology.

Wilson: Much Ado About Calculus.

LaVergne, TN USA
13 December 2009
166829LV00003B/58/A